Sewage and Industrial Effluent Treatment

Sewage and Industrial Effluent Treatment

Second Edition

John Arundel
MSc, CChem, MRSC, FCIWEM

Blackwell
Science

© 2000 by
Blackwell Science Ltd
Editorial Offices:
Osney Mead, Oxford OX2 0EL
25 John Street, London WC1N 2BL
23 Ainslie Place, Edinburgh EH3 6AJ
350 Main Street, Malden
 MA 02148 5018, USA
54 University Street, Carlton
 Victoria 3053, Australia
10, rue Casimir Delavigne
 75006 Paris, France

Other Editorial Offices:

Blackwell Wissenschafts-Verlag GmbH
Kurfürstendamm 57
10707 Berlin, Germany

Blackwell Science KK
MG Kodenmacho Building
7–10 Kodenmacho Nihombashi
Chuo-ku, Tokyo 104, Japan

The right of the Author to be identified as the Author of this Work has been asserted in accordance with the Copyright, Designs and Patents Act 1988.

All rights reserved. No part of this publication may be reproduced, stored in a retrieval system, or transmitted, in any form or by any means, electronic, mechanical, photocopying, recording or otherwise, except as permitted by the UK Copyright, Designs and Patents Act 1988, without the prior permission of the publisher.

First published 1995
This Edition published 2000

Set in 10/12.5 pt Times
by Aarontype Limited
Printed and bound in Great Britain by
MPG Books Ltd, Bodmin, Cornwall

The Blackwell Science logo is a trade mark of Blackwell Science Ltd, registered at the United Kingdom Trade Marks Registry

DISTRIBUTORS

Marston Book Services Ltd
PO Box 269
Abingdon
Oxon OX14 4YN
(*Orders*: Tel: 01235 465500
         Fax: 01235 465555)

USA
Blackwell Science, Inc.
Commerce Place
350 Main Street
Malden, MA 02148 5018
(*Orders*: Tel: 800 759 6102
                781 388 8250
         Fax: 781 388 8255)

Canada
Login Brothers Book Company
324 Saulteaux Crescent
Winnipeg, Manitoba R3J 3T2
(*Orders*: Tel: 204 837-2987
         Fax: 204 837-3116)

Australia
Blackwell Science Pty Ltd
54 University Street
Carlton, Victoria 3053
(*Orders*: Tel: 03 9347 0300
         Fax: 03 9347 5001)

A catalogue record for this title is available from the British Library

ISBN 0-632-05356-9

Library of Congress
Cataloging-in-Publication Data

Arundel, John.
 Sewage and industrial effluent treatment / John Arundel. — 2nd ed.
  p.   cm.
 Includes bibliographical references and index.
  ISBN 0-632-05356-9
  1. Sewage—Purification.   2. Factory and trade waste—Management.
  3. Sewage—Purification—Case studies.
  4. Factory and trade waste—Management Case studies.   I. Title.
 TD745.A78 2000
 628.3—dc21   99-33304
                              CIP

For further information on Blackwell Science, visit our website:
www.blackwell-science.com

Contents

Acknowledgements ix

Common Abbreviations Used in the Water Industry x

Introduction xiii

1 **Preliminary Treatment** 1
  1.1 Introduction 1
  1.2 The importance of preliminary treatment and initial considerations 2
  1.3 Detritus removal 4
  1.4 Grit removal 16
  1.5 Oil and fuel separation 19
  1.6 Conclusions 22
  1.7 References 23
  1.8 Companies and other organisations 24

2 **Suspended Solids Removal by Settlement and Flotation** 25
  2.1 Introduction 25
  2.2 Initial considerations 25
  2.3 Aspects of sedimentation specific to sewage treatment 26
  2.4 Settlement theory 28
  2.5 Settlement tank design parameters 29
  2.6 Settlement tanks and practical operation 33
  2.7 Upward flow clarifiers 41
  2.8 Flotation 43
  2.9 Common problems associated with settlement, clarifiers and DAF 49
  2.10 Conclusions 51
  2.11 Summary table of typical operating parameters and dimensions 52

	2.12	References	52
	2.13	Companies and other organisations	53
3	**Metal Removal Methods**		**54**
	3.1	Introduction	54
	3.2	An overview of effluent treatment options from metal finishing processes	55
	3.3	Effluent toxicity and discharge limits	56
	3.4	Treatment methods for metal-containing effluents	60
	3.5	Plating shop layout and housekeeping	69
	3.6	Metal ion biosorption by microfungi	70
	3.7	Conclusions	71
	3.8	Company	71
4	**Biological Treatment Methods**	**72**
	4.1	Introduction	72
	4.2	Initial considerations for new installations	74
	4.3	Some operating practicalities of biological oxidation	78
	4.4	Biological oxidation microbiology	80
	4.5	Secondary treatment plant types	82
	4.6	Conclusions	103
	4.7	Summary table of typical operating parameters and dimensions	104
	4.8	References	107
	4.9	Companies and other organisations	107
5	**Sludge Disposal and Treatment**	**109**
	5.1	Introduction	109
	5.2	Initial considerations	111
	5.3	Sludge types and characteristics	111
	5.4	An overview of treatment options	112
	5.5	Treatment methods	119
	5.6	Digestion	134
	5.7	Incineration	140
	5.8	Conclusions	147
	5.9	References	149
	5.10	Companies and other organisations	149
6	**Cesspools, Septic Tanks and Small Sewage Treatment Plant**	**151**
	6.1	Introduction	151
	6.2	Choosing a treatment system	153
	6.3	Commissioning new treatment systems	161

	6.4 Treatment systems – layout, operation and maintenance	162
	6.5 Conclusions	178
	6.6 Companies and other organisations	178

7 Developments in Wastewater Treatment — 179
   7.1 Introduction — 179
   7.2 Developments in biological treatment — 180
   7.3 Developments in physical treatment — 195
   7.4 References — 205
   7.5 Companies and other organisations — 206

8 Sampling and Analysis — 207
   8.1 Introduction — 207
   8.2 Sampling: initial considerations — 208
   8.3 Sampling methods — 209
   8.4 Sewage works sampling — 212
   8.5 Sampling industrial sites — 215
   8.6 Sampling equipment and practical techniques — 216
   8.7 Simple sampling statistics — 220
   8.8 Flow measurement methods and equipment — 224
   8.9 Effluent charging methods — 226
   8.10 On-line sensors for continuous monitoring — 228
   8.11 Analysis: initial considerations — 230
   8.12 Conclusions — 238
   8.13 Companies and other organisations — 238

9 Case Histories — 240
   9.1 Introduction — 240
   9.2 Case history no. 1 — 242
   9.3 Case history no. 2 — 246
   9.4 Case history no. 3 — 248
   9.5 Case history no. 4 — 250
   9.6 Conclusions — 254

Appendix 1: Glossary of Terms — 255
Appendix 2: Typical Sewage and Industrial Effluent Plant Layouts — 259
Appendix 3: Imperial/Metric (SI) Conversion Factors Common Wastewater Imperial/Metric (SI) Parameters and Conversion Factors — 264

Index — 267

For Naomi

Acknowledgements

I would like to thank the following for their assistance in the preparation of this book.

The many UK manufacturers of equipment who provided line drawings and operational performance data; each company is listed at the end of the chapter in which it is mentioned.

Former colleagues in Thames Water PLC who reviewed some sections of the text for content, particularly Arthur Paine who reviewed Chapters 1 and 6; also plant supervisors and operators who gave accounts, frequently graphic, of their experiences with some of the equipment.

Thames Water PLC for permission to photograph on-site equipment at a number of STWs; and the Co-op Dairy for permission to photograph one of its treatment plants.

The Chartered Institution of Water and Environmental Management for permission to reproduce a number of tables, diagrams and graphs from papers published in the IWPC and CIWEM Journals.

John Corris of Warren Jones Engineering for permission to reproduce Table 8.1.

Alan Gosling for reviewing Chapter 8 and providing information on metal treatment for Chapter 3.

Mr J. Franklin of BS Floccor Ltd who allowed reproduction of part of his original paper on metal effluent treatment for Chapter 3 and provided information on applications of random plastic media.

Andy Weatherhead of Advanced Drainage Engineers Ltd, Wendover who helped with several photographs in Chapter 6.

British Rail for civilised mobile office and research facilities, very adequate catering and inspirational views.

Lastly my wife Naomi and our children Tom and Victoria, for enduring my sometimes lengthy periods of incarceration in the library at home when I should have been out playing.

Common Abbreviations Used in the Water Industry

AQC   analytical quality control
ATU   allylthiourea
BOD   biochemical oxygen demand
CHP   combined heat and power
COD   chemical oxygen demand
CS   crude sewage
DAF   dissolved air flotation
DM%   dry matter per cent
DO   dissolved oxygen
DS%   dry solids per cent
DWF   dry weather flow
EOD   effective oxygen demand
FE   final effluent
F/M   food/biomass ratio
G & O%   grease and oil per cent
GLC   gas liquid chromatography
grp   glass and reinforced plastic
HMIP   Her Majesty s Inspector of Pollution
HTE   humus tank effluent
IS   intrinsically safe
CIWEM   Chartered Institution of Water and Environmental Management
mg/l   milligrams/litre (equivalent to ppm and μg/ml)
MLSS   mixed liquor suspended solids
NAMAS   National Accreditation Measurement Service
NO_2N   nitrite expressed as nitrogen
NO_3N   nitrate expressed as nitrogen
NRA   National Rivers Authority
O & V%   organic and volatile per cent
pe   population equivalent
PO_4P   phosphate expressed as phosphorus
ppb   parts per billion i.e. 10^{-9}
ppm   parts per million i.e. 10^{-6}
RAS   returned activated sludge
RBC   rotating biological contactor

SAF submerged aerated filters
SAS surplus activated sludge
SCA standing committee of analysts
SCADA supervisory control and data acquisition
SS suspended solids
SSVI stirred sludge volume index
STE sedimentation tank effluent
STP standard temperature and pressure
STW sewage treatment works
SSVI stirred sludge volume index
SVI sludge volume index
TDS% total dry solids per cent
TOC total organic carbon
TWL top water level
UASB upward flow anaerobic sludge blanket system
UFR upward flow rate
μg/ml micrograms/millilitre
VM volatile matter
WL water level

Consent limits in the text e.g. 40/30/10, are given in the conventional order, i.e. SS/BOD/NH$_3$N.

Introduction to the First Edition

When I joined the UK Water Industry in 1972, a wise, if somewhat cynical, old sage observed that the progress of civilisation was directly proportional to the distance mankind placed between itself and the waste it produced. Having historically placed some 'facilities' at a distance. an increasingly crowded world now has to live ever closer to its own waste and effluent. On reflection, I don't ever recall seeing the factory manager's office or the company boardroom anywhere near the effluent treatment plant – yet.

Nevertheless, the future certainty of greater amounts of waste and effluent has served to concentrate minds in the last two decades on how to treat the increasing volumes of waste produced by society, and also how to change things to produce less in the first place. New technology has allowed upgrading and refinement of material expectation, but a lack of political will, personal motivation and sometimes money has kept reform at a slow pace.

But things are now changing. Formidable legislation is in place – and is being slowly enacted in Europe, the USA and the UK – that demands high standards of effluent treatment and waste disposal. The aim of this book is to provide a practical guide for managers, supervisors and operators in both manufacturing industry and the public utility water companies. There is an increasing need for awareness of available choices and options in equipment and techniques to improve the standard of effluent discharges from factories and treatment plant and thus compliance with the new, tougher discharge limits that the regulatory bodies now seek Coincidentally, other savings and recycling may become possible.

I hope that designers, engineers and consultants will also find this book a useful, practical reference source as to the likely performance of the treatment options available for both domestic sewage and industrial wastewater: the text to each chapter attempts to provide an objective assessment of the available systems. More detailed performance is contained in the references and manufacturers' literature. Technical and theoretical considerations are explored, and practical solutions reinforced by a number of case histories, providing parallels with which readers may identify, and which they can then adapt to their individual circumstances.

However, it is worth remembering that the performance of every biological wastewater treatment plant is unique, as reflected by the individual composition of the incoming wastewater. While most operate within an accepted range, examples exist which, despite the application of standard design criteria, fail to achieve the desired performance levels. Conversely, there are others which considerably out-perform expectation in adverse environments.

I hope those with only limited knowledge of wastewater treatment will find the book an 'easy read' and, in view of the reference nature to which it might be put, key points are repeated in the text to chapters and sections. Readers unfamiliar with abbreviations and the jargon of water treatment are referred to Appendix 1, where a reasonably comprehensive list has been compiled.

With over 100 years of worldwide water supply and treatment experience, UK water industry and engineering design philosophy have been characterised by conservatism and lengthy life expectations for civil structures. This approach has also governed those compiling the specification of mechanical and electrical equipment on water treatment works and for underground services. Designs are therefore robust and reliable and survive long periods of continuous operation in aggressive environments. Refurbishment and upgrading is possible long after many 'domestic' items would have been scrapped.

Long life is expected because of the high capital cost of installation and replacement. It is a sobering thought that many cities in Britain rely on an infrastructure of underground pipes laid in the last century, many kilometres of which are still in excellent condition. There are therefore thousands of working examples of the types of equipment described in this book, some of them 30–40 years old and located at sites that predate the twentieth century.

Many new technological innovations in wastewater treatment have been employed in the last decade, and the privatisation of the UK water industry has spawned a manufacturing growth market for 'package' and high-rate treatment plant, providing quick solutions to problems that were often the result of previous neglect. Such equipment often demands sophisticated continuous control, but there are advantages.

Package plant offers low civil construction costs, installation time and expense, the possibility of containerising the equipment and modifying treatment or increasing capacity easily and relatively cheaply. Industry has used package plant successfully for many years, but has not always demanded exacting performance. However, even plant of some vintage, if structurally sound, responds to some 'tuning'.

The water industry has moved cautiously down this path in the UK, particularly since privatisation, where the cost benefits and short construction times have assisted with rapid placement of a number of schemes for sewage treatment in response to EC Directives. This approach represents a substantial change from the high cost, land consumptive but very enduring conventional sewage treatment facility operated conservatively and with some spare capacity. Recently, the UK Water PLCs, keen to reduce their own capital expenditure on new works, have been starting to apply pressure on industry to clean up its effluent independently and reduce volumes discharged, or face rapidly escalating discharge costs.

Newcomers to treatment plant and equipment from industry, faced with installing some treatment facilities for the first time, should be aware that although the capital costs are high, so too is manufactured equipment quality, and the industry is well supported by a number of long-established manufacturers whose products are in worldwide use. They must also consider carefully the vices and virtues of both conventional and package treatment approaches, and in particular the long-term energy requirements of some of the newer ideas. Manufacturers are usually very willing to conduct free on-site trials of their products to determine treatment potential of wastewater and sludges. This book has two main themes:

(1) The need to make the most of existing treatment facilities, and, where these are non-existent or inadequate, to seek simple, reliable solutions.

I am no enthusiast for overt sophistication, and while the microprocessor has its place in monitoring, data acquisition and control, particularly on remote installations, it remains no substitute in the fluid industrial environment for motivated and experienced personnel who are better able to react and adapt. Staff training is thus fundamental to satisfactory results.

Industry is about making things for profit, and wastewater treatment about operational reliability and design adequacy. They are complementary activities, but in order to be effective partners, do require an overview of the whole production and discharge cycle on site.

(2) The need to produce less waste requiring treatment in the first place.

Recovery and recycling have been fashionable before; the UK drought of 1976 served as a foretaste of the consequences of long-term water shortage and limited supply. Old habits returned afterwards, with just as much product ending up on the floor, just as many hoses left running all day, heat losses warming the neighbours and unselective dumping in the ever-convenient skip. In a recent estimate, UK industry threw away £360 million worth of raw materials per annum in such activities, and it has become obvious recently that when world economies take a downturn, such waste converts profits into losses. Those with good environmental management in operation always have the competitive edge.

Make time now for an objective look round your plant and a look at how materials and water might be used more economically and profitably. At the same time, seek advice from the regulatory bodies about likely future standards for effluent discharges and put aside some capital to make the improvements now, rather than when forced to do so.

John Arundel
Winchester House
36 Winchester Road
Walton-on-Thames
Surrey KT12 2RH
Tel: 01932 225568
Fax: 01932 252482

Introduction to the Second Edition

Since publication of the first edition, my consultancy work has included appraisals and surveys of a number of small sewage treatment works for clients. It came as a surprise to find out that in the UK, approaching one million people rely on private facilities in lieu of main drainage. Their experiences mirror the quality of the installations and I have seen both ends of the spectrum. Chapter 6 is thus an appropriate addition to the book and I hope will prove useful to plant owners, particularly unwilling ones.

The methods reviewed in 'New Developments in Wastewater Treatment' have enjoyed mixed fortunes in the last five years.

Membrane technology has gathered momentum along with UV disinfection and an increasing number of sea outfall effluent discharges have both. Nutrient removal of both nitrogen and phosphorus is often now incorporated in extensions or retrofit schemes, particularly at activated sludge or extended aeration plants.

Deep shaft appears to have waned in popularity and anaerobic treatment systems remain more popular in Europe and the USA than the UK by a factor of twenty times or more.

The potential for odour production and its control at treatment plant has become a high priority in many planning applications and public enquiries though much of the technology, although effective, has little long term documentation or pedigree. The placing of the whole works underground in sensitive environments is also providing engineering challenges unknown ten years earlier. However, the Water companies are also beginning to wake up to the energy costs of running intensive treatment plant and audits can make unattractive reading.

I hope that this book continues to be a useful guide to operators, students and practical men of science.

JA

Chapter 1
Preliminary Treatment

1.1 Introduction

This chapter describes the unsophisticated but essential methods of removing detritus and grit from sewage and other waste waters. Detritus, as used in waste water engineering, is a collective team for gross solids, rag, plastics and general trash, which is separately collected before grit removal at a conventional sewage treatment works (STW); the methods and equipment described in this chapter are equally useful for STW and industrial waste water treatment.

Grit comprises inorganic material ranging from fine sand/silt particles about 0.1 mm in diameter to material over $1\,cm^3$ in volume and of irregular shape. Surface water run-off and general erosion of the sewerage system are the usual sources.

Screens to remove detritus vary from coarse (more than 50 mm between the bars) to fine (1–2 mm), and are usually electrically driven. There are a number of manufacturers and designs, all with their own characteristics.

Simple filter sacks have recently become available, and provide an effective solution for screening out plastics, which are a particular contemporary problem in that they break through screens to primary sedimentation tanks and secondary treatment, and can even penetrate to the works outfall. Wire baskets and grids more suited to the cleaner factory effluent can often be made in-house out of cheap weld mesh.

Grit is highly abrasive and often blocks pipework; separation is based on reducing flow rates to assist gravity settlement, although the material at a sewage works is too variable in grain size and organic content to have any value, and is dumped or disposed to landfill.

Physical methods of separating oil, petrol and fuel/water mixtures are also described and all are simple means of avoiding a highly visible and unattractive type of pollution, one which is immediately obvious to the general public and the regulating bodies.

Industrial effluents containing oil emulsions require more specialised treatment by cracking with acid, or solvent extraction. They are often successfully handled in admixture with other organic wastes by one of the biological processes outlined in Chapter 4.

1.2 The importance of preliminary treatment and initial considerations

Preliminary treatment is essential at STWs as the water companies have very limited control over the amount of general rubbish that can end up in sewers, and even less over the grit and inorganic particles arising from erosion of pipes and joints and from roads and paths where the sewerage system is combined.

Mechanical coarse screens are also located at some pumping stations, particularly on combined (i.e. foul and surface) water sewerage systems, and in drainage systems with a number of pumping stages, those suffering considerable infiltration or where rag and large items are often found in the sewage. These factors are always worth considering in any new scheme or rehabilitation. Pump blockage and damage in the sewerage system is expensive on parts and manpower, particularly overtime, and there is the associated risk of flooding and pollution. By contrast, even a poorly run industrial site is in a much better position to prevent gross material from entering liquid effluent flow by simple housekeeping, and strategically placed grids and wire baskets.

From an economical viewpoint, the methods and formulae used by the water companies to charge for trade effluent always include a 'strength' factor, and Chapter 7 covers these aspects in some detail. Some form of screening will often reduce the overall effluent strength by sifting out larger solids before they break down and require removal by sedimentation or biological means.

If an effluent plant is operating on the limit of organic loading and performance, fine screening can tip the balance away from the overloaded state, or at least give some breathing space. Hydraulic loading and head loss across screens are attendant factors to consider in this situation.

Satisfactory operation of all subsequent stages of a treatment plant often critically depends on correct installation, operation and regular maintenance of screens, traps and interceptors. A number of serious problems are likely to develop from neglecting these devices, including excessive wear and breakdown of pumps, blockage and erosion of pipework, and increased maintenance requirements to equipment throughout the plant. There is then the associated chance of the works effluent failing discharge Consent limits.

Accumulations of detritus reduce the efficiency of biological treatment by generating dead volume in tanks, and may impair settlement and sludge digestion rates, the latter often caused by rag 'balling'. Detritus building up on weirs causes erratic flow measurement, and influences plant operation by incorrect chemical conditioning dosing or incorrect recirculation rates. If pumped, the latter can induce wide flow variation through the plant which may also encourage departures of effluent standard outside Consent limits.

Screenings in sewage have changed substantially in recent years, consisting of far more non-biodegradable plastics material of fine gauge and fewer cellulose-based products (see CIRIA/WRC (1984)). Plastic is persistent, and many coarse screen designs retain little of it, while some older designs of macerators and shredders cut it into ribbons which are even more difficult to retain.

Substantial growth has taken place in the fine screen market during the 1980s, as plastics are now prohibited substances in the aquatic environment and must not be discharged. The trend has been to wash and remove screenings, followed by landfill disposal or incineration. Dewatering and compaction devices frequently complement a package screen installation, and substantially reduce screenings disposal volumes. There has been a parallel trend in Western Europe for the *per capita* volume of screenings to increase, and $0.3\,\mathrm{m^3/day}$ per 1000 population is often quoted as an average value.

The waste disposal unit under the kitchen sink is the most likely culprit causing this trend, and whilst mainly increasing the biological and sludge loadings to the works, also generates fibrous material likely to be trapped and effectively blind fine screens below 5 mm mesh size. The water industry might justifiably promote a policy of actively discouraging these devices which are estimated to increase the *per capita* solids loadings to the sewer by 25% per day.

Screenings and detritus, particularly of sewage origin, and oil/water mixtures are very noticeable near beaches or inland surface waters, and frequently attract attention by the public and regulatory bodies out of proportion to the actual pollution level because they are considered aesthetically revolting. The function of screening sewage discharges to sea is therefore largely concerned with reducing the identifiable nature of the solids, as much of the material tends to float.

There are a number of sewage discharges to sea in the UK and many worldwide which receive no treatment other than preliminary solids screening. Screening is too coarse at many of them, reliance being placed on dilution, mechanical breakdown by waves and tide and distance from the shore. Many storm overflows are not screened at all. Guidelines issued by Her Majesty's Inspectorate of Pollution (HMIP) as a letter to the UK Water Authorities in 1988 suggest that a mesh size greater than 6 mm is unlikely to achieve adequate performance of long sea outfalls which are the subject of EEC Directive 76/160/EEC (1976) Concerning the Quality of Bathing Waters: '...that all flows should be subject to fine screening...' The HMIP letter also discusses the extent of screening in relation to the magnitude of storm flows. Other aspects to consider for marine discharges include the general requirements of Directive 91/271/EEC (1991), the Council Directive Concerning Urban Wastewater Treatment, which dictates that preliminary screening must be continuously effective, and adds the requirement for primary settlement and possibly secondary aeration to many of these discharges.

In such cases, preliminary treatment might help to reduce the bacterial numbers in subsequent discharges by solids removal, also useful in complying with Directive 76/160/EEC (1976). 'Might' is the operative word, as the action of many screens is to break down solids as well as remove larger material so that there is no difference between the organic load measured as Biochemical Oxygen Demand (BOD) either side of some screen designs.

Discharges of sewage to the marine and estuarine environment are currently receiving particular attention in Europe. A useful reference paper on screen performance, particularly in relation to the plastics problem, is provided by Thomas et al. (1989).

Conventional grit removal on sewage works takes place after screening to remove detritus. Some grit will therefore be trapped in fine screens and will act as an abrasive on the screen surfaces and the cutters of macerators or comminutors. It is desirable, therefore, to specify hardened or stainless materials for screens.

Unremoved grit tends to accumulate on sewage works in sludge digestion and aeration tanks, reducing effective capacity, circulation and sludge retention times. This has several negative effects. These include reducing gas production rates and energy availability in boilers where external and expensive energy 'topping up' may become necessary; lowering the quality of the sludge and liquors produced, affecting works performance; and generating heavy maintenance costs on mixers and pumps arising from abrasive wear to seals, bearings and metalwork.

Oil discharges to treatment plant make an unsightly mess of tank and channel walls which are labour intensive to clean; the oil also coats filter surfaces, reducing oxygen transfer and performance. There is frequently an odour problem from volatile components or if the material emulsifies and turns rancid, and many mineral oils contain toxic organics and heavy metals after use. Trade effluent Consents specifically exclude flammable and explosive fuels from discharges to the foul sewer, and contemporary limits of 500 mg/l for oil are typical from garage forecourt interceptors and car washes.

It is worth noting that oil discharges are the easiest to trace back to source either in rivers or sewers, by virtue of their persistence and ready observation, and the elegantly conclusive analytical techniques which are able to uniquely 'fingerprint' oil; thus a number of cases brought to court by the National Rivers Authority (NRA, 1991) in the UK have resulted in successful prosecutions and the imposition of spectacular fines. There is thus practical, commercial and legal sense behind some of the humble devices now described.

1.3 Detritus removal

1.3.1 Simple manual screens and traps

The simplest form of screen needs consist of nothing more than a wire mesh basket through which effluent can flow, the mesh size being dictated by the nature of the material being handled. A plating shop, for instance, might expect to capture nuts and bolts, metal swarf and pieces of plastic from jigs, locating the traps at the outlet of each plating tank to prevent the need to drain rinses and dragouts for a vital piece of lost work.

Even for similar effluents which contain low suspended solids, a mesh trap somewhere in the drainage system is essential, and should be given occasional but regular inspection. Workers in production areas are just as likely to lose rings, loose change, watches, hairclips and even wallets as those in the offices!

Food and dairy processing often produces fat and grease particles which can block pumps and produce odour problems in tanks. An effective wire mesh screen or series of baskets at the inlet to treatment plant will remove this, but such items need regular inspection and cleaning as they often rapidly become blocked by sticky particles, and a mechanically raked fine screen may be more suitable here.

A simple screen consisting of a disposable sack made with 3 mm terylene mesh is available from Copa Products Ltd, one of several suppliers; it has been widely used on smaller sewage treatment plants to prevent the discharge of plastics, rag and grease balls from storm tank overflows, and in the dosing chambers of percolating filters, where pricking out the filter arms to remove such debris is a regular maintenance task. The sacks are attached to weirs on tanks or in the dosing syphon chamber, and have a self-cleaning action, whilst also breaking down faecal matter. Figure 1.1 shows a typical installation. A version for marine discharges is supplied with an agitation system to assist the

Figure 1.1 Copa sack installation on a weir.

solids break-up process. They could be equally effectively incorporated in a gully on the factory floor, or at the overflow of drag-outs and settlement tanks.

A recent innovation at sewage treatment plant is to skim the surface of secondary settlement tanks with submersible pump sets to capture and disperse scum. The device can be modified to remove plastic that has passed unscathed through activated sludge treatment, collecting the material in a mesh sack suspended over the stilling box. This is a successful, if 'last ditch', method of preventing these materials entering the watercourse. Figure 1.2 shows this arrangement, a final settlement tank and the volume collected in less than 24 hours.

Industrial production areas generating gross solids must be fitted with floor gullies covered by gratings with 12–20 mm square slots which are screwed or bolted down, and certainly not easily lifted. This removes the temptation to hose everything on the floor down the drain at the end of a shift or working day – or even worse, to clear a temporary flood by lifting the gratings.

There is no finer way of blocking the drainage system of the factory than by hosing down powdery or granular material that then coagulates and sets overnight in pipe bends and dead ends. The sewerage and sewage operations sections of the water companies also take a dim view of lumps of concrete, meat hooks, bone, plastic swarf and pieces of wood in the public foul sewer; ultimately these can jam pumps or screens with expensive results, and may

Figure 1.2 Surface pump and plastics skimming trap.

cause flooding for which the water companies become liable if it causes pollution. They therefore tend to devote some effort to recovering costs from the original owners of the material and to prosecuting guilty parties for any pollution indirectly caused.

A simple and very fine rundown screen (0.25–0.75 mm aperture) (one such is manufactured by Environmental Engineering Ltd) is useful for continuously screening solids from agricultural waste containing straw, vegetable washings, peelings and similar food processing waste, and textile washings containing lint, fibres and very fine grit. This device (Figures 1.3 and 1.4) has no moving parts, solids being continuously collected and often quite dry.

Some headloss is inevitable with any screening device, and in the factory

Figure 1.3 Rundown screen. (Courtesy: Environmental Engineering Ltd.)

8 *Sewage and Industrial Effluent Treatment*

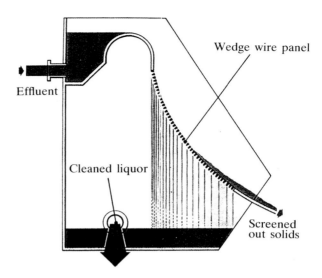

Figure 1.4 Cross-section through a rundown screen. (Courtesy: Environmental Engineering Ltd.)

environment, the placing of screens in the flow should be checked carefully to prevent any backing-up or flooding. Draining effluent to a sump, followed by pumping to a higher level for treatment is a solution demanding highly reliable pumps, and is best considered in combination with other treatment apart from simple screening, in view of the energy costs.

1.3.2 Mechanical screens

There has been considerable advancement in the range and performance of mechanical screens in the last decade, giving the designer a wide choice to suit a number of environments. Development has been driven by both economical and environmental considerations, and treatment process changes. Sewage works have been progressively automated – entirely demanned on some of the smaller plants – and now rely on the travelling gang for periodic attention and emergency call-out via a telemetry link. Some of the older preliminary treatment equipment at the works inlet requires near-constant attention, particularly during storm conditions when a lot of debris is flushed from the sewers.

With few exceptions, screens are driven by a motor/gearbox combination, typically 2 hp which is often enclosed within the body of the screen. UK Health and Safety requirements dictate enclosure of moving parts.

A number of factors need consideration when choosing screens. Choice of type and mesh size will be dictated by a number of on-site and location factors. It is essential that a reasonably comprehensive survey of the size range of detritus is carried out, as sewage often has a unique local quality such as a preponderance of rag, excessive fat or farm materials like straw. Some of the more recent sewage treatment package plants need no primary sedimentation tanks and are of sufficient capacity to avoid the need for flow balancing, but these do benefit from fine screening and grit removal at the inlet.

Extended aeration in oxidation ditches, for example, uses a fairly slow circulation rate of mixed liquor in which grit will readily settle and rag bind into lengths of rope if not previously removed at the works inlet. Reductions in aeration tank capacity of 25% have been experienced after five years' operation of these systems at sites in the UK connected to uncombined sewerage systems, i.e. from which surface water is apparently excluded. Effective fine screening and grit removal are thus particularly important for extended aeration systems.

Breakthrough of plastics from the older types of coarse bar screen (more than 50 mm mesh) poses particular problems, and has necessitated the fitment of wire baskets or surface skimming pumps, as already mentioned, to final tanks and effluent chambers on many works to comply with the prohibition on plastics being discharged to the aquatic environment.

It is obvious that coarse bar screens (more than 50 mm mesh) are no longer appropriate choices for marine discharges and neither is any form of screenings maceration which returns the material to the flow, albeit as smaller particles. Screens of nominal mesh size 15–18 mm only retain about 50% of the screenings load and are equally ineffective in marine applications, where much finer screening (to 6 mm or less) is now required so that the resultant liquid discharge contains no readily identifiable solids.

Most marine discharges are remote and sites must now contain adequate storage facilities for collected screenings and grit, and be completely automated. Thomas *et al.* (1989) provide a useful assessment of a number of screen types that could be used at sea outfalls.

The practice of macerating screenings and returning them to the flow has declined recently in favour of dewatering and bagging, with, at some installations, on-site incineration. Incineration of a material containing a lot of plastics places stringent emission and combustion control requirements on the process and may be uneconomic on a small scale. Screenings incineration plant is the subject of control under the Environmental Protection Act 1990, Part 1.

High reliability is expected of mechanical screens and one practical consideration is their elevated and exposed position on many works inlets. Freezing up in cold weather is a frequent operation problem, encouraged by the damp atmosphere directly above channels. Some form of thermostatically controlled frost protection including trace heating and insulating washwater supplies, drive chains, gearboxes, hoppers and compactors and condensation heaters in motors should be considered essential in the specification for sites likely to experience frost or freezing fog.

Headloss through coarse screens is typically 50 mm, and may be as much as 150 mm across a fine 4 mm screen, similar to the comminutor that was popular in the 1970s. Choice may therefore be influenced by slight gradients in approach channels and a shallow slope to the works site.

The traditional coarse bar screen is shown in Figure 1.5. The coarse bar screen is an essential component of the inlet works at many larger sewage works connected to a sewerage system where the sewers are more than 300 mm in diameter, combined sewerage systems and those with high infiltration. The primary function of this type of screen is to protect pumps and other equipment in subsequent processes. Some of the largest UK works are frequent recipients of items of clothing, timber and metalwork like bicycle frames brought down by storm flows. The bars are typically 20–50 mm apart, and collect rag, large debris and fat balls quite effectively, trapping smaller plastics floating at right angles as the screen blocks.

Operation is automatic, and usually initiated by the differential head that builds up either side of the bars as they become occluded. This is detected by float switches or ultrasonic liquid level detectors, and most systems can be adjusted within a 25–75 mm head difference range. Timed screen operation is also possible, but not appropriate if flows and loads vary widely, as they do at most sewage treatment works.

Most manufacturers can supply bar screens for a standard range of channel

Figure 1.5 Coarse bar screen.

widths, and different bars spacings are available, down to 12 mm. Other variations on the same theme include stepped bar screens (manufacturers of these include Vickerys Ltd.) some of which are continuously cleaned by a polypropylene brush which will effectively reduce the amount of fat clinging to the bars.

The drive consists of a motor of about 2 hp, gearbox(es) and control equipment linked to level detection. The drive is often a chain on enclosed screens or external oscillating arms and must contain a shear pin.

1.3.3 Fine screens

Most fine screens are cylindrical, being slowly rotated by an electrical drive similar to bar screens. Diameters vary, material is usually of stainless steel, and the whole structure should be of sufficient strength to withstand operators using manual methods to remove colloidal matter. Fine screens can be used to relieve overloaded primary sedimentation tanks by removing some of the potentially settleable solids, and may reduce the biological load to the plant, but depending on the size range of debris arriving on site may blind very quickly. Again, a site survey of screenings is essential before making a decision.

Fine screening has been used in the food industry for some time; and it has become widely used in sewage screening recently to remove plastics and reduce particle size. Mesh size covers the range 0.25–15 mm. An aperture size of 6 mm has been adopted in the UK as the maximum acceptable for Consent applications dealing with marine discharges designed to protect bathing use.

When used to screen sewage, fine screens capture large quantities of debris with a high faecal content; if they are replacing coarser bar screens, this must be allowed for by extra storage capacity and washing/dewatering facilities. Figure 1.6 shows an installation at a sewage works where there is no primary sedimentation, the main treatment system being an oxidation ditch.

Most fine screen designs are variations on a theme. The drum screen admits flow from the outside, trapping debris on the outside surface to be scraped or pressure-washed off. Effluent discharges axially from inside the drum. Cup screens work in the reverse manner and solids are trapped on the inside drum surface to be washed off into a channel inside the drum. The microstrainer, used as a tertiary treatment system on sewage works and an initial coarse screen of reservoir water for potable supply is an example. It is essential that the screen is continuously cleaned.

Other variations contain nylon brushes and polypropylene scrapers, and are particularly appropriate for fat-bound materials, e.g. pet foods and abattoir waste where screen blinding occurs. A 40% reduction in suspended solids (SS) and a 30% reduction in BOD is claimed by Lockertex Ltd, one of the manufacturers of drum screens.

Figure 1.6 Cylindrical fine drum screen.

Mesh blinding is a frequent problem with fine screens; the apertures may become completely blocked by fine strands of solids weaving in and out of them. This is termed 'hairpinning'. Sewage containing a high rag and hair content is a typical culprit. Doubling the drum rotation speed, significantly increasing wash water pressure and mechanical brushing have all been used with success to combat mesh blinding. Substantially increasing the screen mesh thickness also makes a significant improvement. Some claims have been made that plastic mesh is less prone to the problem than metal. The theoretical design aspects of cup screens have been well reviewed by, for example, Stamper and Graham (1981).

1.3.4 Screenings and detritus disposal

Screenings collected by any of the above methods are gathered into a trough or screw conveyor flushed with sewage or works effluent, or a hopper from which excess liquid can drain, and are further treated by either maceration, dewatering and compaction, or incineration.

Maceration

In this method the material is passed through an in-line disintegrator, which reduces the size of the detritus, and then returned to the flow. This was a popular technique in the 1960s and 1970s; the theory was that reduced material would settle better in subsequent primary sedimentation and cause less damage en route. The presence of increasing amounts of plastic in sewage defeats the object, and the method is not suitable for any secondary treatment system (for example, extended aeration) without primary sedimentation or grit separation. The macerator unit itself is a high-maintenance item and can be readily damaged or jammed by bits of wood and rag balls.

A contemporary version consists of a macipump discharging screenings from a collection hopper into a centrifugal dewaterer. A very consistent product is obtained, the device dealing well with plastics. It is often teamed with a fine drum screen to collect the screenings initially, with grids to prevent the passage of any large items (over 50 cm). Figure 1.7 outlines the installation of four units at a sewage works with a dry weather flow of 12 000 m^3/day, and reducing the material collected by two fine 6 mm screens. The product discharged to a skip below the macipumps and dewaterer (Figure 1.8), is less than 15% organic matter. Suppliers of these units include Haigh Engineering Co Ltd.

Dewatering and compaction

This is a favoured method, as it reduces the volume for eventual disposal. Screenings are moved by a screw auger to a hopper, loaded into a second conveyor and either compacted by a ram or driven up an inclined plane by auger, the water draining from the lower end. The discharge is often continuous and into a trailer. A volume reduction of 75% is possible with a compactor. A number of manufacturers supply such equipment, often as a complete package with a screen set. In this state, screenings are a potential odour problem and quickly start to decompose. On many small sewage works, screenings are buried on-site and quickly rot down.

The author saw an interesting experiment in the 1970s: screenings were bagged in black airtight sacks; on inspection after a year they had completely composted to a fine dark material, and the majority of the plastics either powdered or disintegrated.

Handling larger volumes presents a disposal problem. Bagging and landfilling has a transport cost and available sites that will accept what is defined as a hazardous waste may be limited. Nevertheless, this is a common disposal route, and co-disposal is usually the cheaper off-site option. On remote sites, some form of bagging is essential if vermin are not to be attracted coupled with a regular removal routine.

Several practicalities arise with bagging. The plastic bags used should be strong enough to withstand manhandling without tearing, and small enough to lift and load by hand (25 kg). A few drainage holes in the sack base will allow any further water to drain away, as after a week of storage, some natural dewatering takes place. The bagging unit needs partial enclosure or shelter, as bags are prone to freezing in northern climates due to the exposed areas of the conveyor and compactor and the elevated aspect of many works inlets.

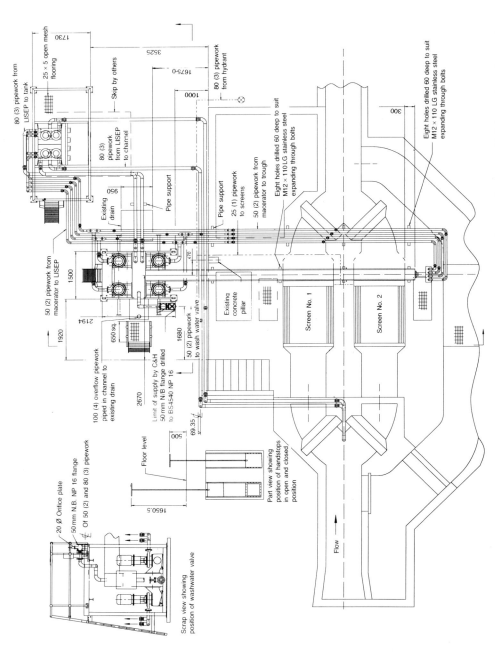

Figure 1.7 Macipump installation. (All dimensions are in mm with inches in brackets.) (Courtesy: Haigh Engineering Ltd.)

Preliminary Treatment 15

Figure 1.8 Macipump screenings product.

Incineration

This ultimate disposal route is attractive because if sufficiently dewatered to more than about 35% dry matter, screenings will burn autothermically. Some practical experiences are presented by Pace and Price (1982).

Stringent emission standards have been placed on incinerators in Part 1 of the Environmental Protection Act (1990). Contemporary screenings containing a lot of plastic will require incineration at temperatures above 1000°C to prevent dioxin formation; these will generate hydrogen chloride gas and bring about the need for flue gas scrubbing.

Incineration costs are therefore high compared to those of landfill, but if carefully engineered incineration does provide on-site disposal and the prospect of heat recovery. The low ash volume generated would be acceptable for landfilling, but some leachate trials for metals, particularly zinc and cadmium, would be advisable.

1.4 Grit removal

Grit removal is the second essential preliminary stage of sewage treatment, but it is comparatively rare in industrial effluent treatment installations as a separate process, being frequently combined with primary settlement. Grit arises in sewage from the erosion of pipework and joints, and infiltration into leaking sewers through sand and gravel strata.

In larger drainage systems where the surface, storm and foul sewerage systems are combined, large volumes of eroded road surface material and general dirt from vehicles add to the flow, and skip loads can be collected at a large works following heavy rain. Ongoing renovation of old sewerage systems involving the separation of surface and foul water provides welcome improvements and will assist economic use and design of capacity at treatment plant, not only in the appropriate sizing of grit and detritus removal facilities but in coping with a more normal flow variation in wet weather. It is important therefore to size grit removal facilities in relation to the type of sewerage system, and to allow leeway for known infiltration rates and probable increase as the pipework degenerates.

Grit removal is essential at sewage treatment works for two main reasons:

(1) to reduce blockage, wear and erosion of pump bodies, bearings, stators and rotors, and pipework, environments where it will act in the manner of a fluidised grinding paste;
(2) to prevent the reduction in available operating capacity that occurs when grit is deposited in sludge tanks and digestors, and in oxidation ditches which are surface aerated and where the flow velocity around the system is inadequate to keep inorganic material in suspension.

The effectiveness of the grit removal stage directly dictates the frequency with which these tanks and digestors require draining and the grit laboriously and expensively dug or jetted out. A 25% reduction in tank capacity arising from grit build-up is not uncommon in less than 10 years of continuous digestor operation, and creates operation problems and the potential for works performance sliding below discharge Consent limits.

As a cause of slow deterioration in works effluent quality, this factor should not be overlooked because grit removal efficiency varies widely with flow and is never 100% for any contemporary equipment. The dead volume of a tank arisng from retained grit can be fairly simply measured by comparing actual versus theoretical retention times using a suitable tracer such as sodium chloride.

Prior to grit removal, most waste water treatment plant inlet channels are made deliberately narrow, but most are of sufficient capacity to cope with an order of magnitude change in received water flow rate, in order to maintain velocity and retain all materials in suspension.

Conventionally, grit is removed after detritus has been taken out, and a common method is to reduce the flow rate of incoming waste water to a linear surface velocity of less than 0.3 m/s by widening the incoming channel to form a chamber with inlet and outlet baffles. A scraper, top-driven from an overhead bridge, slowly rotates, sweeping the grit into a collecting hopper.

From here, collected material could be pumped or drained out, but as the hopper is normally flooded and in constant use, the grit is usually removed by pushing it up an inclined channel with an oscillating arm with attached scraper blades. Figure 1.9 shows a cross-section through such a detritor commonly found in the water industry. Materials ranging in size between 0.5 mm and 15 mm (silt to pea shingle) are effectively removed because, being denser than water, they rapidly fall out of suspension. Periodic drainage of this device is necessary to clear grit collecting at the corners of the grit chamber, and which the scraper will miss.

Two other methods of grit removal that are often installed at sewage treatment works are illustrated in Figures 1.10 and 1.11. The first method comprises a couple of channels in parallel, V-shaped so that as flow increases, so too does available cross-sectional area. The sloping side walls also assist in grit deposition at the base of the channels, from which the grit is sucked by vacuum pump, and ultimately discharged to a dry hopper. This is a land-consumptive device, and has been superseded by the second method, a vortex design offered by several manufacturers including Tuke & Bell Ltd. and Promech Environmental Ltd.

Flow enters and is swirled round in a cone-shaped structure. Mechanical stirring separates organic and inorganic solids, the latter falling to the base hopper. After washing, the grit is removed by airlift pump or screw. The device is compact: 95% removal of 100 mesh particles is claimed, and particles down to 0.15 mm may be retained at normal flows.

The variable grit removal efficiency of these devices at higher flow rates is inevitable and has already been mentioned. Since flow rate is highly variable on many sewage works, grit removal equipment represents a design compromise.

A consensus of water industry experience is that the detritor design shown in Figure 1.9 removes about 85% of grit entering the works; analysis of this material shows it to be 60%−70% inorganic.

The vortex design is much more dependent on the maintenance of air blowing or stirring rates to keep organic and inorganic particles separated, and can quickly fill up unnoticed. Less grit is then trapped. Nevertheless, large volumes of grit are removed by the devices described, particularly during high flows in winter, when road and concrete surfaces and pipework break up from mechanical and frost damage, and silted storm drains and

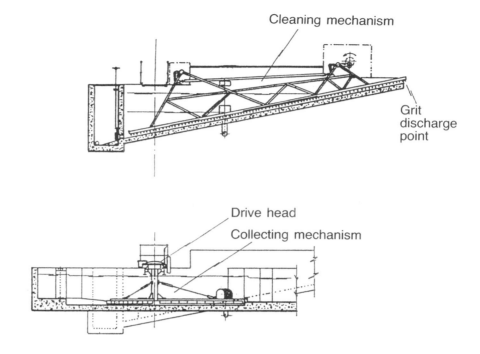
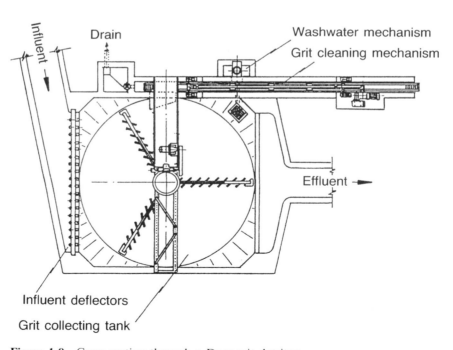

Figure 1.9 Cross-section through a Dorr grit detritor.

Figure 1.10 V-grit channels.

overflows are flushed out. Quantities of sand are collected at works near to the coast, arising from infiltration through sandy strata and washing clothes and towels taken onto beaches.

Although some of the above-mentioned devices wash the grit during extraction, it is rarely more than 70% inorganic when of sewage origin and always contains entrained organic solids and food particles like peas and beans, and scraps of plastic that escaped screening. The moisture content varies widely between 15% and 70%. The material is thus valueless, and on most small sewage works sites it is dumped in a convenient hole. An odour nuisance can be generated in summer if fresh material is not buried. Larger sites must observe the Duty of Care Section 34, Part 1 of the Environmental Protection Act 1990 (that such material must be disposed of in a licensed tip).

Industrial users who may find the vortex type of grit separator useful for trapping aggregates like grit and sand will be able to recycle uncontaminated material, but they must check the device regularly, and in particular the nozzles on air-lift pumps.

1.5 Oil and fuel separation

Facilities are not normally provided at sewage works for the removal of floating oils because they are specifically excluded or, typically, limited to

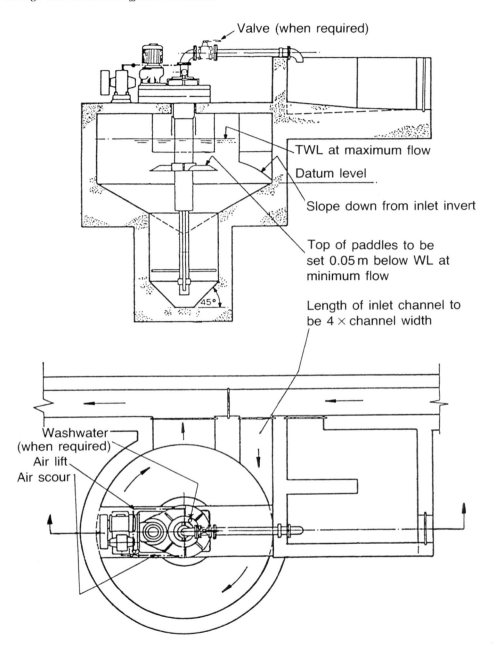

Figure 1.11 Cross-section through a vortex grit trap.

levels of 500 mg/l in industrial Discharge Consents to the foul sewer. The arrival of diesel, gas oil or petrol at a works indicates a spillage or accident requiring immediate investigation and possibly flow diversion to storm tanks. Fortunately, this type of pollution is easily spotted and traced to source by the judicious lifting of manhole covers throughout the sewerage system, the material always leaving traces on pipes and benching in chambers.

The NRA report (1991) revealed that nearly 25% of substantiated incidences of river pollution involved oil. The material is highly visible, and thus all industrial discharges to surface waters where there is the slightest chance of oil contamination must consider one of the removal techniques below.

Although oil storage tanks must be installed in bunds, some bunds are of inadequate total capacity when collected rainwater adds to the volume, and in no circumstances should the bund have drain holes! A small hand pump should be used to empty the bund contents, which may require treatment if contaminated.

1.5.1 Oil/water separation methods

Classically, oil/water mixtures can be separated in tanks or interceptors, the water fraction being abstracted towards the tank base for discharge, and the oil skimmed or pumped off. Providing flow rates allow quiescent conditions, and baffles are placed at inlet and outlet, this simple method works well.

Large separators used by oil refineries normally achieve a 500 mg/l standard for discharge to estuaries, although partially soluble components such as phenols and benzene derivatives are discharged and the water often has a characteristic odour. By 1995, the Paris/Oslo Convention will have tightened discharge limits to less than 10 mg/l total hydrocarbons and this simple method will not be acceptable. Some form of biological treatment, probably activated sludge with phosphorus and nitrogen nutrient addition, will be required to achieve these limits at large oil refineries and depots.

On the more modest level of the garage forecourt or small factory, a simple structure is adequate to achieve the 500 mg/l trade effluent discharge Consent limit; this structure normally has three chambers in series. Construction is often below ground and of brick. A lot of grit and silt also enters the first chamber when draining paved areas, and the position of the interconnecting pipes is as important as regular cleaning out.

Ventilation is essential where petrol spillages may also be contained in the device. Packaged versions are also available in glass reinforced plastic (GRP) from Klargester Environmental Products Ltd (Figure 1.12) and are claimed to reduce hydrocarbon levels to 5 mg/l through the inclusion of a coalescer unit. This standard is increasingly sought by regulatory authorities where the discharge is to a river or stream.

Tilted plate separation is another useful technique for treating oily water containing 10 – 50 mg/l oil, and offers a considerable space saving over large

22 Sewage and Industrial Effluent Treatment

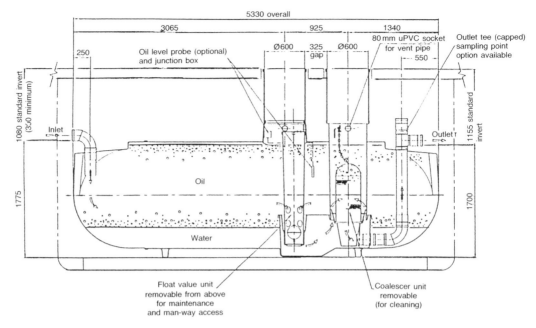

Figure 1.12 Cross-section through a package oil/petrol interceptor. (All dimensions are in mm.) (Courtesy: Klargester Environmental Engineering Ltd.)

settling tanks. Figure 1.13 shows a modular and expandable layout, with closely spaced parallel corrugated plates inclined at 45° or more. Oil globules move upwards and coalesce, form larger droplets and rise to the surface, while sediment slides down the plates. Laminar flow conditions are essential through the plates. More information is available from Environmental Engineering Ltd, and this manufacturer also supplies separators capable of treating the first polluting flush of storm flows from hard standing areas, which are often heavily contaminated with oil products.

Oils, fats and grease from food industry processing are better separated by dissolved air flotation, and this technique is described in Chapter 2.

1.6 Conclusions

Many of the operating problems to be found on sewage works and industrial effluent plants arise from blockages and loss of treatment capacity caused by detritus, grit and fat deposits. Some form of preliminary treatment is essential. Attention to the capacity and detail of this relatively simple aspect of wastewater treatment will reap dividends in terms of operating reliability, maintenance, effluent quality and the manpower needs of the whole treatment site.

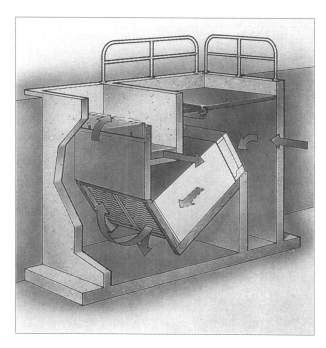

Figure 1.13 Tilted plate oil separator.

1.7 References

CIRIA/WRC (1984) *Screenings and grit in sewage – removal, treatment and disposal*. Preliminary report of the Construction Industry Research and Information Association and the Water Research Centre, Technical Note 119 Medmenham, Buckinghamshire.

Directive 76/160/EEC (1976) Council Directive Concerning the Quality of Bathing Water. *Official Journal L31/1*. HMSO, London.

Directive 91/271/EEC (1991) Council Directive Concerning Urban Wastewater Treatment. *Official Journal L135/40*. HMSO, London.

Environmental Protection Act (1990) *Part 1 – Integrated pollution control and air pollution control by local authorities*. HMSO, London.

Her Majesty's Inspectorate of Pollution (1988) *Guidance on long sea outfalls – Copa Consent applications*. Letter to managing directors of regional water authorities. HMSO, London.

NRA (1991) *Water pollution incidents in England and Wales*. National Rivers Authority, HMSO, London.

Pace, D.W. & Price, G.J. (1982) A solution to screenings problems by dewatering and incineration. *Journal of the Institution of Water Pollution Control*, **81** (3).

Stamper, A. & Graham, N.J.D. (1981) Towards improving the specific rating of cup screens in sewage flows. *Journal of the Institution of Water Pollution Control*, **80** (1).

Thomas, D.K., Brown, S.J. and Harrington, D.W. (1989) Screenings at marine outfall works. *Journal of the Institution of Water and Environmental Management*, **3** (6).

1.8 Companies and other organisations

Copa Products Ltd, Copaclear House, Crest Industrial Estate, Pattenden Lane, Marden, Tonbridge, Kent TN12 9QJ.
Environmental Engineering Ltd, Little London, Spalding, Lincolnshire PE11 2UE.
Haigh Engineering Co Ltd, Ross-on-Wye, Herefordshire HR9 5NG.
Klargester Environmental Products Ltd, Aston Clinton, Aylesbury, Buckinghamshire.
Lockertex Ltd, PO Box 161, Church Street, Warrington, Cheshire WA1 2SU.
Promech Environmental Ltd, 75 Manchester Road, Congleton, Cheshire CW12 2HT.
Tuke & Bell Ltd, Beacon Street, Litchfield, Staffordshire WS13 7BB.
Vickerys Ltd, 53 Norman Road, Greenwich, London SE10 9QJ.

Chapter 2
Suspended Solids Removal by Settlement and Flotation

2.1 Introduction

This chapter examines available techniques for suspended solids removal, either by settlement (sedimentation) or flotation. Settlement is by far the more common technique in wastewater treatment; it involves little mechanical equipment and is the more stable to operate. There are some situations where flotation is the better choice however, or where settlement might not even work; the removal of fats and proteins from food industry effluent, or fine inorganic particulates produced during mining are classic examples.

Clarifiers generating an intermediate level sludge blanket are widely used in the water supply industry and are successfully used for fibrous solids removal in, for example, the paper industry. Simple clarifiers placed on top of a conventional settlement tank and using wedge wire mesh or gravel media, are useful for upgrading solids removal performance, or where secondary treatment is overloaded. Very high quality effluents with negligible solids content are possible after passage through upward flow clarifiers.

At the 'low-tech' end, lagoons have much to recommend them where there is plenty of space and possible odour generation is not a concern.

2.2 Initial considerations

Solids removal dramatically reduces the 'strength' of most effluents and is widely practised. It is one of the simplest and most cost-effective ways of reducing trade effluent treatment charges.

Industrial effluents discharging to estuaries and coastal waters are frequently only treated by settlement. The sludge is often capable of reuse, and if inorganic, the liquid supernatent may be virtually unpolluted and similarly recycled as second-grade washwater.

As a general rule, effective settlement of most effluents reduces the solids by at least 50%, with a proportional reduction in oxygen demand or 'strength' measured as Biochemical Oxygen Demand (BOD) or Chemical Oxygen Demand (COD) and ranging between 30 and 50%. Tanks that are lightly hydraulically loaded and receiving effluent with coarse or well-mineralised solids may achieve 90% removal, as might a clarifier generating a sludge blanket. A 10% reduction of bacterial numbers during primary sedimentation is usually assumed in design calculations for sewage treatment.

Both settlement and flotation is easy in theory; in practice, some attention to details pays dividends in effecting satisfactory separation of solids and water.

The density difference between most waterborne solids that will form sludges and water is marginal; most organic solids lie in the range 1.03–1.10. Any major disturbance to the system – changing flow rates, turbulence, thermal stratification or sudden withdrawal of sludge or liquid – that upsets quiescent conditions within the tank, affects settlement.

Settlement tanks can have a useful balancing action in themselves by buffering sudden changes of organic and hydraulic load applied to secondary treatment, and, if carefully sized, iron out flow variations through the rest of the plant. To effect maximum settlement, however, some form of flow balancing before settlement tanks may be necessary, particularly in industrial environments where 'wash-ups' occur at the end of each shift or production batch.

Addition of chemicals to assist settlement by coagulating particles or chemical precipitation is essential for treating a variety of industrial wastewaters, particularly those containing metals, and is reviewed more fully in Chapter 3.

The Victorian method of adding lime as a slurry to sewage has been revitalised recently and is worthy of mention as it can be a useful physico-chemical treatment for an overloaded works, reduce bacterial numbers in discharges to sea where the high pH will be of less concern and assist in phosphate removal prior to discharge to inland waters where eutrophic conditions exist. The use of both lime and aluminium sulphate in an upward flow clarifier are described by Jones *et al.* (1991). The increased sludge volume produced by such treatment (25%–30% is typical) is a major operational and cost consideration.

To summarise, sedimentation in some form is thus essential for most systems, and must be operated effectively to prevent deleterious effects on later treatment stages. A number of industrial effluents can be successfully treated to acceptable discharge standards by well designed settlement facilities with regular desludging, preventing the need for further and more elaborate secondary treatment. This possibility should always be examined in any review of treatment options.

2.3 Aspects of sedimentation specific to sewage treatment

Sedimentation is the first and primary treatment stage of conventional sewage treatment for biofilter and activated sludge systems, although some extended aeration plant dispense with any form of primary treatment and treat crude sewage after screening and grit removal. The resultant raw sludge has a strong and unpleasant odour, requires some form of treatment before disposal, and is typically 4%−5% dry matter (DM).

Primary tanks are normally desludged at intervals of between 8 and 24 hours, which allows some consolidation to occur and produces a thicker sludge, whilst avoiding septicity and anaerobic conditions which might cause odour problems and rising sludge. Individual circumstances will apply to each site, and are best determined by local knowledge or past experience as to the 'freshness' of the sewage arriving for treatment.

Many European sewage discharges to estuaries and the sea are at present treated only by primary sedimentation, and the resultant settled sewage discharge may be acceptable in achieving quality and bacterial standards with adequate dilution and outfall placement.

2.3.1 Secondary sludges

Secondary settlement during sewage or industrial effluent treatment then follows any form of biological aeration or filtration, and performs the vital function of separating oxidised humus or activated sludge to produce an effluent low in solids, usually between 5 and 30 mg/l, which in most cases is of adequate quality for discharge to a watercourse.

Activated sludge is returned to the aeration plant to maintain the solids levels, while humus sludges from filters are mixed with primary sludges before further dewatering or more elaborate treatment such as anaerobic digestion.

As many organic secondary sludges are difficult to dewater, some initial in-tank thickening is highly desirable, but this requires careful control. An effective method is to control the sludge withdrawal rate by optical sensors suspended in the tank. A light beam measures the position of the sludge/water interface, which can be set by experiment but is normally at one-third to a half tank depth from top water level. The interface is normally sharp for secondary sludges, and the detector can then be made to control desludging valves automatically.

Such a system is widely used in activated and extended aeration sewage treatment plant, where incoming solids of 2500−6000 mg/l can be usefully thickened up to 20 000 mg/l (2%), giving a head-start to subsequent dewatering and reducing volumes to pump and handle off-site. Secondary sludges have an earthy and not unpleasant odour and are lighter and more floc-like than raw sludge. They are universally more difficult to dewater than primary sludges; Chapter 5 discusses this problem more fully.

Secondary sludges are biologically active, often with a high respiration rate and a nitrogen content of 3%−8%. Much of the soluble nitrogen is present as nitrate and will readily denitrify, the released nitrogen gas then causing the light sludge flocs to float (referred to as a tank inversion or rising sludge) and be lost with the effluent. This is potentially a significant cause of pollution from sewage works discharges in warm weather (temperatures above about

15°C). It is an essential requirement, therefore, that secondary settlement tanks are desludged either continuously or at frequent intervals, by contrast with primary tanks. This leads to variations in mechanical scraper design and the slope of the tank base.

Particularly demanding discharge consents may then dictate tertiary treatment to remove solids and BOD by a further 50% and the options here include lagoons, sand filters or microstrainers.

2.4 Settlement theory

The basic theory is simple. Effluent enters a tank, either in batches or continuously, where a combination of low flow rate and time allows suspended solids particles to descend by gravity and form an underlying sludge. The clarified top liquid layer, the supernatant, overflows or is decanted as effluent for discharge or further treatment.

Particle settlement rate is dependent on three main factors:

(1) particle size;
(2) the density difference between solid and liquid phases;
(3) the liquid viscosity.

The relationship is expressed mathematically in Stokes' equation:

$$U = \frac{gD^2 (d_1 - d_2)}{18u}$$

Where:

U is the terminal velocity of the particle.
g is the acceleration due to gravity.
D is the particle diameter.
d_1 is the particle density.
d_2 is the fluid density.
u is the fluid viscosity.

The settlement rate thus varies directly with the square of the particle diameter and, if the fluid is water and $d_2 = 1$, with the particle density.

Stokes' Law applies to regular, spherical discrete particles, not shapes associated with most sludge solids and particularly fibres where settlement rates will be significantly less than theoretical. However, the relationship with the square of the diameter provides the basis for encouraging flocculation and agglomeration of particles. Addition of chemicals such as polyelectrolytes, which reduce the electrostatic forces of repulsion between suspended matter, are added to assist flocculation, but are not essential.

Slow mechanical stirring and quiescent conditions will encourage flocculation by particle collision and attachment, and are just as effective with biological and highly organic sludges. Similarly, Stokes' Law accounts for the superior settling characteristics of more dense, granular primary sludges by relating particle density to settlement rate.

2.5 Settlement tank design parameters

There are three basic considerations.

2.5.1 Upward flow rate.

As solid particles are gravitating downwards against an upward flow of liquor, the upward flow rate (UFR) is of prime importance. This is expressed in metres per hour, and is also known as the surface loading, or surface overflow rate, defined by the equation:

$$\frac{\text{Effluent feed rate/hour}(m^3/h)}{\text{Tank surface area }(m^2)}$$

The water industry designs conservatively for this parameter and tanks are sized to give UFRs of $0.5-1.0\,m^3/m^2/h$, which will cope with the lightest of activated sludges; 1.2 m/h is an often-quoted design parameter for 'average' conditions. Heavier industrial solids may well settle out at flow rates above $2\,m^3/m^2/h$, but sudden flow variations are likely to cause settlement problems at high UFRs resulting in less than 50% solids removal, which is the performance level to aim for. An effluent high in dissolved solids or soluble BOD is unlikely to settle out effectively at the higher rates unless chemically dosed before entering the settlement stage.

A simple way of increasing the weir surface area and effective length — and this is widely practised in sewage treatment — is to fit V-notch weirs around the tank. As the flow rate increases, the liquid level rises further up the V-notch and so has a larger area to flow over. Tanks which suffer a wide flow variation can benefit settlement by fitting a number of such weirs. Figure 2.1 shows an installation where two are fitted, the effluent flowing into a separate channel suspended within the tank.

2.5.2 Retention time

The second design consideration is to give sufficient time for the solid particles to reach the tank base or underlying sludge layer, i.e. the retention or detention time of the effluent within the tank. This is simply defined as:

$$\frac{\text{Tank volume }(m^3)}{\text{Effluent feed rate }(m^3/h)}$$

When sizing settlement tanks, the retention time at the maximum likely flow rate for an anticipated 50%+ solids removal is the important calculation. A

Figure 2.1 Double-sided V-notch weirs on settlement tank.

realistic appraisal of flow variations is essential, as there may be flow rate changes of ten times in some industrial situations, or where the effluent is entirely pumped to the tanks. In practice, retention times vary quite widely in wastewater treatment. As little as 2 hours is possible even with organic raw sludges and particularly heavy inorganic solids, but it is normally regarded as a practical minimum. Five to eight hours is a typical range in most primary sedimentation tanks at sewage works, the primary tanks of small works often being desludged at intervals longer than 24 hours without any problems, providing the sewage is not septic on arrival. Secondary settlement tanks are normally sized for a retention time of 4–5 hours, allowing sufficient time for the sludge to settle but for little consolidation other than some agglomeration to occur, so avoiding denitrification and floating material.

Some industrial operators will have tanks where the retention time is measured in days, and providing the sludge is inorganic or does not decompose or react readily, this will allow compaction to occur and give a useful reduction in sludge volume. Related issues dictating against very long retention times include temperature, both ambient and that of the effluent, with the attendant risk of decomposition, and the thickness or dry matter content of the settled sludge that can be handled easily. Most sludges will not flow readily out of valves or down pipes if more than 10% DM, and pumping costs can be high at these levels.

Long retention times and sludge compaction may therefore bring greater operational problems than a regular desludging regime which produces a greater volume of thinner sludge.

2.5.3 Maximum solids loading (MSL)

The third design parameter relates to the likely suspended solids levels in the effluent entering the tank, from which it is possible to calculate the MSL, i.e. the mass of solids per unit area per unit time that can be applied to that tank before solids flow upwards and over the weirs with the effluent.

This parameter has been extensively researched in relation to the design of secondary settlement tanks on activated sludge plants, but applies equally to settlement of metal hydroxide sludges. WRC (1975) describes how this parameter can be measured, and is essential reading for detailed study.

In summary, the solids loading is made up of two components: that due to the settling of sludge flocs under gravity, and the sludge withdrawal rate from the tank. The MSL can be obtained by plotting solids loading in kg/m^2/h against solids concentration in g/l. A graph similar to Figure 2.2 is obtained where:

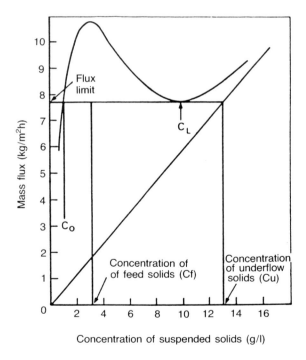

Figure 2.2 Relationship between solids loading and solids concentration below the critical rate of sludge withdrawal. (Courtesy: CIWEM.)

Cf is the feed solids concentration (3.5 g/l).

Cu is the sludge withdrawn concentration in g/l.

The lowest point in the curve C_L represents the maximum solids loading that can be applied before the solids rise and leave with the effluent, and also the critical value for sludge withdrawal. Under normal operating conditions, therefore, the rate of sludge withdrawal determines the solids loading that can be applied and should be above the critical value.

The settling characteristics of the sludge are equally important and can be measured as follows. A sample of effluent is slowly stirred in a cylinder of about 4 l capacity, and after 0.5 h the volume of settled sludge in millilitres is noted. This is divided by the solids concentration to give the stirred specific volume (SSV) in ml/g. By convention, the SSV is determined at a solids concentration of 3.5 g/l, this being a typical suspended solids value of influent activated sludge in secondary settlement tanks.

The SSV, sludge withdrawal rate, solids loading and flow rate through the tank are all related by a nomograph (Figure 2.3). On the predicted side, Qu is the volumetric rate of sludge return and A the total surface area of the settlement tank. The solids loading in $kg/m^2/h$ is found by using a ruler to connect the SSV

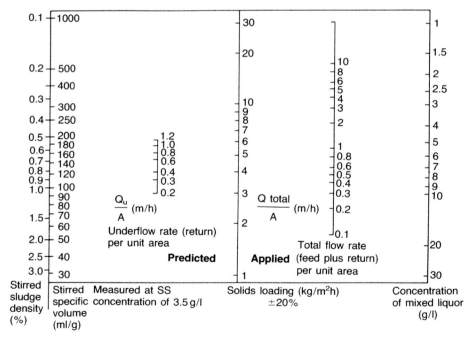

Fig. 2.3 Nomograph of predicted and applied solids loading. (Courtesy: CIWEM.)

and Qu/A values. On the applied side of the nomograph, the total flow rate Qu includes the volume of influent flow to the tank and the sludge return. Knowing the MLSS in g/l allows the solids loading to be similarly obtained. In practice, maximum solids loadings on activated sludge plants are about 20% lower than those predicted by the nomograph, and some conservative sizing of tanks is therefore wise. SSV values that might be obtained are dependent on the sludge age in the plant, but 80–140 ml/g at 3.5 g/l concentration is a reasonable range, and anything over 150 ml/g, particularly in an extended aeration plant, should be carefully monitored.

2.6 Settlement tanks and practical operation

2.6.1 Lagoons

The simplest form of settlement tank need consist of nothing more than an excavated area with earth walls. If an industrial site has the luxury of spare or derelict land, this option should be given more than a cursory glance in assessing treatment options to reduce suspended solids levels. The effluent quality will be much improved, and the frequency and cost of tankering sludge off-site dramatically reduced in many cases. Limiting factors are ground conditions, the nature of the effluent, sludge and related odours.

Clearly, a high water table or the possibility of polluting streams close to the surface will rule out lagoons unless effectively lined by butyl sheet or clay. This requirement may make construction uneconomic or impractical if the lagoon is ultimately to be bulldozed, or the sludge recovered by mechanical excavation.

As most lagoons are usually shallow and of considerable surface area, high annual rainfall which is only partially offset by evaporative losses may use up much of the available volume, and is a relevant factor in tropical areas. Lagoons have been successfully used to capture inorganic solids from the paper and cement industries, and in farming, where the odour potential of decomposing organic sludge is accepted, or remote from concentrated housing.

On the farm, in order to avoid gross hydraulic overloading during storms it is vital that the minimum amount of rainwater – particularly that collected from roofs – mixes with yard and animal pen run-off draining to a lagoon. A number of serious pollution incidents have occurred in recent years from lagoon banks bursting or over-topping, or the discharge being very high in solids and organic content because of minimal retention times.

If odour is not a concern, lagoons receiving organic sludges will act effectively as cold, slow digestors and may be ploughed up eventually. This practice is widely used in tropical rural areas as part of a complete sewage treatment system, where a number of lagoons are used in cascade and by rotation, the digested sludge being of considerable agricultural value and the effluent well purified and often low in pathogens. The high ambient temperatures are a benefit here, providing the sludge is kept covered by liquid.

Most lagoons are designed with retention times measured in days or weeks rather than hours to effect maximum solids removals, and a simple inlet distribution system at several points will help in spreading the sludge load evenly. The long retention time offers significant buffering capacity to attenuate periods of poor effluent quality, and the shallowness will assist mixing, neutralising reactions and coagulation in the industrial situation.

Ultimately, if the sludge is to be pumped out, 10% DM is a safe maximum for most materials, otherwise some form of dragline or excavator will be necessary. Cleaning out should be necessary only once every 1–2 years, and it could be carried out at much longer intervals than this. Hire of or investment in mechanical plant will be necessary, with associated transport costs. The recovery of the sludge for reuse or sale may be quite practical with careful working. Typical lagoon performance would be a sludge volume reduction by compaction or digestion of over 35%, and 40%–75% solids removal.

In sewage treatment, UK practice has been to use lagoons as a tertiary or polishing stage because they are simple, reasonably low in cost to construct and repair, and consume no power. In this application, their shallow construction assists ultra-violet (UV) penetration, and bacterial kill rates. If bacteriological standards are to be applied to sewage works effluents in the future, this will be a much cheaper option, given the space, than any method of artificial disinfection.

2.6.2 Rectangular settlement tanks

Uncommon now except on the largest sewage treatment plants, these tanks are easier to construct than the circular type and are quite common in a number of industries where the need to desludge the tank regularly is less important. A typical late-Victorian sewage works would have been blessed with these structures, constructed with elaborate brickwork and requiring manual desludging, neither a pleasant or particularly safe work activity.

Modern installations should be built with a ratio of width to length in the range 1:2 to 1:4, baffled inlet and a floor sloping slightly towards the effluent discharge end, to encourage sludge to gently roll into a well or discharge trough. A V-shaped floor (as seen in cross-section) with a central channel is an alternative arrangement.

Mechanical scrapers suitable for these tanks are more complex, requiring traversing and lifting functions. Most installations have no scraper set and rely on gravity to collect and withdraw sludge. If the tank is less than 2 m deep, any baffles or scum boards fitted should only penetrate the top water level by 10–15 cm to avoid turbulence and disturbing the underlying sludge layer by scouring. The baffle performs the important task of preventing influent liquid streaming across the top surface carrying floating debris to the effluent channel, and effects some mixing to reduce thermal stratification. In many situations, the incoming liquid is warmer than the tank contents.

2.6.3 Circular settlement tanks

This has become the standard design on sewage treatment plant, where construction in concrete has become a well practised art. Circular in plan, with a ratio of diameter to depth usually within the range 3:1–8:1, the tank base slopes from the periphery inwards by as much as 30°, encouraging the sludge to a central collection hopper, from where it can be withdrawn by pump or hydrostatic head.

Flat-floored circular settlement tanks are cheaper to construct, but uncommon, needing effective scraping. Most tanks are fitted with a mechanical scraping set. The effluent enters via a central pipe into a stilling box, flowing first down and outwards and then upwards to discharge over a peripheral weir. The tank is normally buried in the ground to top water level, requiring considerable initial excavation; this policy is dictated on many sites by the need to maintain a fall through the whole works so that all the flow proceeds by gravity, and costly intermediate pumping is avoided.

The long-term pumping costs of any elevated tank should always be considered in the budget for plant operation, as a poorly-laid-out site, where the disposition of the main components, including balancing and settlement tanks, is illogical or does not use a slope to advantage, will incur a constant energy cost. The construction quality, particularly of buried concrete tanks, is very enduring. They will be fit for decades of service, outliving the mechanical scraping equipment and other metalwork several times over.

High groundwater levels must be allowed for in buried tanks by providing flap valves in the tank to allow hydrostatic pressure relief when the tank is drained down. More than one sewage works in the world can boast a tank that has risen out of the ground by a measurable amount, and often wrecked pipework and electrical connections on route! Prefabricated settlement tanks made from grp or mild steel are more appropriate in many industrial areas, and can be installed at ground level very quickly. Among suppliers of these, Environmental Engineering Ltd is typical, offering an installation where a wedge wire clarifier can be fitted to the top and a sludge blanket generated. This additional feature acts as a filter, and 95% success rates in solids removals are claimed. Normal desludging is effected through a large gate valve at the base of the tank as required.

Two important considerations for above-ground tanks within the factory area are point loadings and sizes. A relatively small tank, 6 m in diameter and 4 m high excluding the hopper base, will weigh 115 t when full. It is equally important to have adequate settlement capacity, pump sizes and inlet baffling to avoid disturbance. In many situations, a daily fill, settle and desludge system of working can be used. It may be possible to dovetail shift-working and effluent production so that the tank serves two different production areas and handles different effluents. Solids can then be kept separate and may be recovered and recycled.

Different effluents that may react and beneficially coagulate are best treated in an initial balancing or mixing facility upstream of the settlement tank. Circular tanks settling organic sludges requiring regular or constant removal are usually fitted with some form of scraper mechanism. There are a number of options and design varies with the type of sludge.

Fixed bridge scrapers have a stationary bridge spanning the tank, with the scraper supported and driven from the centre by a motor of 0.5–2.0 hp. Final drive is through a double reduction worm gearbox, and a special spur gearbox

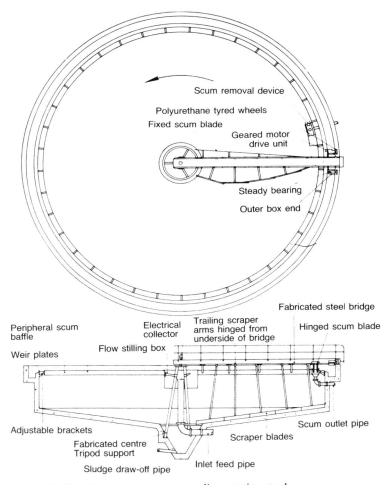

Figure 2.4A Half-bridge scraper set on sedimentation tank.

Figure 2.4B Drive motor and gearbox – scraper bridge.

with a substantial thrust bearing supporting the suspended parts. The central drive shaft is located on the tank base in a spigot. About 20 m in diameter is the largest practical size.

Half-bridge scraper sets are the more common, as they can be installed satisfactorily on tanks up to 45 m in diameter, and there has been a trend towards larger settlement tank sizes. One end of the bridge is supported and pivots at the tank centre, while the other end is mounted on an end carriage that runs round the tank outer wall on a wheel (or rail in earlier designs). Contemporary installations have diminutive drive motors of about 300 W driving 1000:1 ratio gearboxes directly or via a chain. Figure 2.4A shows this arrangement and Figure 2.4B the drive unit.

As the bridge rotates, some allowance must be made for minor variation in the tank structure and wheel wear, so central pivot bearings are designed to take a small amount of vertical movement.

Power collection is via a slip ring collector at the tank centre, fed from either an overhead cable supported by catenary wire or by cable ducted up through the tank floor and stilling box. A shear pin is an essential component of the drive mechanism of either scraper type.

The type and design of scraper blade and scraper rotational speed are dictated by sludge type. Primary raw and heavy industrial semi-organic sludges are usually moved by blades hinged at the tank base and supported by rigid arms. A speed of 2.5 m/minute at the blade tip is possible without disturbance.

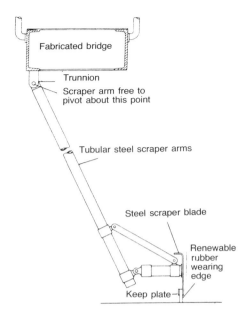

Figure 2.5 Trailing pivoted scraper arm – sedimentation tank.

Secondary humus and activated sludge are normally scraped with a blade hinged at the upper part of the arm; another hinge at the blade end (Figure 2.5) can be provided and is probably the ideal trailing scraper, reducing blade wear and trailing weight, and giving considerable compensation for minor irregularities in the screed of the tank floor. A rotation speed of about 1.2 m/minute is a safe maximum for lighter, flocculant sludges. In both cases, the scraper blades are arranged in the form of a continuous volute (Figure 2.6), so that sludge is removed to the central well with every revolution around the tank.

Activated sludge tanks with a 30° floor slope often desludge satisfactorily with a simple chain scraper to disturb the sludge and encourage it to roll down to the central well. Tanks built with flat floors and settling light, secondary or flocculant sludges prone to denitrification are often fitted with V-shaped scrapers and individual sludge draw-off tubes. Figure 2.7 shows this arrangement, the sludge being removed by hydrostatic pressure. An air lift pump can also be used, the compressed air serving to scour the tubes clean and make them less prone to blockage.

A minimum life of ten years should be expected from the mechanical equipment described above.

A continuous scum baffle around the tank periphery, and about 30 cm inboard of the weir, is an essential feature for all tanks where there is the slightest chance of floating material. This has become particularly important

Suspended Solids Removal by Settlement and Flotation 39

Figure 2.6 Continuous scraper blades – sedimentation tank.

on secondary tanks, where penetration of plastics through conventional screens often results in their appearance at many sewage works after extensive secondary treatment.

On primary sedimentation tanks at sewage works, the scum baffle will prevent most floating grease particles and screenings from fouling the weirs and blocking filter arms, unless radical changes in upward flow rate sweep the material under the board. A board depth of 15 cm into the liquid is recommended.

A scum collection box, slightly raised above top water level, can be provided, a small blade sweeping material collected against the scum board into the box with each bridge rotation. A hinged flap on the trailing side of the box depresses briefly as the blade passes, admitting tank water and flushing the scum into a discharge pipe (Figure 2.8).

To counter the floating plastic screenings problem on secondary tanks mentioned in Chapter 1 (see also Figure 1.2), submersible pumps have been used to skim the top water surface and collect the material for discharge through a filter sack suspended over the central stilling box; this has proved successful. Figure 1.2 shows this operation on a settlement tank at an activated sludge plant. The volume of material in the sack is indicative of 24 hours collection and also reveals the magnitude of this problem in sewage treatment for materials that are now prohibited in effluent discharges to watercourses.

Desludging of mechanically scraped tanks settling primary, raw sludge is usually carried out every 5–12 hours, exceptions being dependent on the

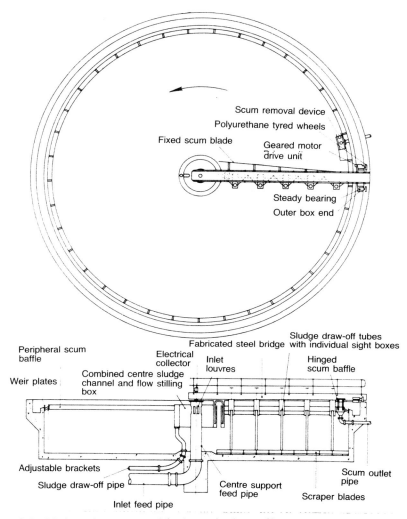

Figure 2.7 V-shaped scraper and hydrostatic draw-off.

'freshness' of the influent. A 4%–6% DM sludge is obtained. Periodic desludging allows some compaction and saves on storage and transportation costs.

The tendency for active secondary sludges, high in nitrogen, to float rather than settle has been mentioned, and a constant or very regular desludging regime is essential for humus or secondary settlement tanks, with sufficient storage to accommodate quantities of sludge rarely more than 4% DM. This is particularly relevant in Mediterranean and tropical climates where raw sludges quickly turn septic (if the sewage arriving at the works is not already septic) and high temperatures cause rapid denitrification of secondary sludges.

Figure 2.8 Scum box mechanism.

The volumes of sludge produced by biotreatment systems are very variable and the factors influencing this aspect of works operation are outlined further in Chapter 4.

2.7 Upward flow clarifiers

At its simplest, a wedge wire panel can be inserted into a settlement tank about 250 mm below top water level and (by the selection of a suitable aperture size) it will act as a fine filter. This is quite often used as an alternative to building more tanks in overload situations; settlement performance can be improved due to two effects:

(1) UFRs vary quite widely through tanks, particularly those of the free-standing package variety; they are highest near the weir. Inserting a panel will equalise the hydraulic load across the tank and give better settlement generally, and especially of the larger irregular particles where the flow rate varies widely.
(2) Small particles or 'fines' whose settlement velocity is close to, or even less than, the upward flow rate will be trapped and tend to coagulate within the wire mesh, forming a blanket which will itself act as a filter. This effect can result in spectacular solids removals of over 95% for industrial effluents containing fibrous material. Figure 2.9 shows a cross-section through an installation supplied by Environmental Engineering Ltd.

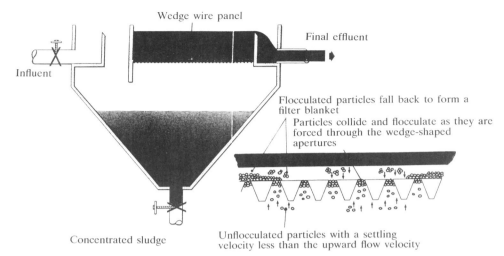

Figure 2.9 Cross-section through a wire mesh clarifier. (Courtesy: Environmental Engineering Ltd.)

The conventional clarifier, widely used for the primary treatment of surface waters during potable water production, generates a sludge blanket midway up a conical bottomed tank. Effluent is admitted through one or preferably several baffled inlets towards the tank base, and passes up through the sludge blanket, where a combination of physical trapping and coagulation ensures the high removal of solids customary for this system.

Aluminium sulphate, polyelectrolytes and granular activated carbon are all potential additives to assist flocculation, and doses can be determined experimentally. Fibrous solids will form a stable blanket without chemical aids, and the clarifier is widely used in the paper industry, where fibre recovery is economically essential.

The sludge is collected by cones suspended under water at the desired top sludge position. A maintained blanket thickness of 0.3–0.9 m is typical with a rise rate of 3–5 m/h. Automation is completed by optical sludge blanket detectors, which control sludge withdrawal rates and maintain the blanket position by adjusting the effluent weir height or incoming flow rate.

A crucial aspect of clarifier operation is that large flow variations do not occur suddenly through the tank; upstream flow balancing or careful distribution where there are several tanks is therefore essential. As a guide, there is a 3:1 ratio between minimum flow that will just maintain the blanket, and the maximum which causes it to rise and expand suddenly.

The UFR for clarifiers is often quoted in the same range as that for settlement tanks: $0.5–1.0 \, m^3/m^2/h$. There are plenty of practical examples of clarifiers working periodically outside this range. The crucial point for those working at high rates is for the blanket detection system and sludge withdrawal facilities to be smart and quick enough to respond to the flow rate changes and so prevent a blanket loss.

In dimension, most clarifiers are built to a diameter/depth ratio between 0.5:1 and 0.8:1, making a fairly deep or tall structure compared to footprint size. Figure 2.10 is a cross-section through a clarifier designed for surface water treatment.

In summary, clarifiers provide effective solids removal, but need quick and responsive control and stable flows to maximise potential.

The upgrading of settlement tanks by lamella plate separators is receiving renewed attention at present and is outlined in Chapter 6.

2.8 Flotation

Flotation is an appropriate method of solids removal if the density difference between the solids and water is marginal, or the solids have a high fat or oil content. There are two methods:

(1) The dispersed air method, which introduces the air through porous media or by an impellor, producing large air bubbles approximately 1000 μm in diameter. Frothing agents are added. The technique is widely used for ore concentration.
(2) Dissolved air flotation (DAF), which is much the commoner technique.

The main requirement of the DAF process is to attach gaseous microbubbles to the effluent feed solids, reduce their apparent density to below that of water, and provide still conditions in the separation zone.

The dissolved air method relies on the observation that the solubility of air in water increases with pressure. A compressor is used to inject air into a proportion of the treated effluent, which is then recycled under pressure at about 3 atmospheres. When released into the flotation tank at atmospheric pressure through nozzles, a cloud of microbubbles is produced, typically 70–90 μm in diameter. These attach to solids, encourage coalescence and cause the solids to rise rapidly to the surface. In contemporary recirculation systems, the air compressor, a pressure pump and the air injector operate in conjunction with a 1–3 minute retention vessel and at operating pressures of 3–10 bar.

A constant pressure of air is maintained to the nozzles by discharging the compressor through a pressure cylinder acting as a reservoir, also useful if brief power interruption is common on-site. Tanks are usually 2–4 m high, and the design of the air injection nozzles is fairly critical, controlling flow rate but without inducing turbulence which might shear relatively delicate solid flocs. UFRs of 3–7 $m^3/m^2/h$ are often quoted. In potable water treatment, rates up to 12 $m^3/m^2/h$ could be realised; conversely, sludge thickening is often achieved successfully at only 1–2 $m^3/m^2/h$. Solids loading rates of 10 kg/m^2/h are realistic for organic, oxidised sludges. The volume of the recycled

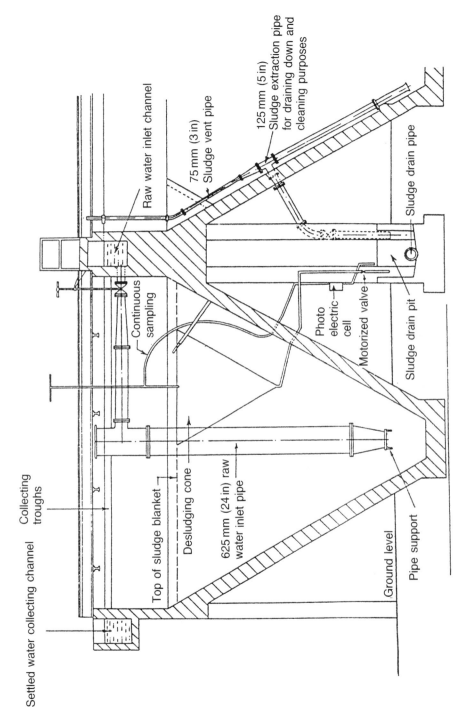

Figure 2.10 Cross-section through an upward flow clarifier.

effluent can vary between 15% and 100% of the total flow. Satisfactory performance primarily depends on the ratio of the volume of air (A) to the mass of solids (S). A/S is typically 0.005:1 to 0.006:1.

Some form of flocculant aid is often essential for satisfactory coagulation; polyelectrolytes are widely used in conjunction with ferrous sulphate and aluminium sulphate, and are added in a flash mixer or stirred dosing tank before air sparging. Concentration is not critical, except on economic grounds. The likelihood of the ferrous sulphate being reduced to the sulphide in high BOD wastewaters after the DAF plant and generating hydrogen sulphide is a potential disadvantage. Likewise, some polyelectrolytes have a narrow pH range for effective flocculating action and careful pH control is often an additional requirement for successful DAF operation.

Solids accumulate on the water surface and are continuously scraped off into a trough with a slightly raised lip, by a rotating bar where the tank is circular. A typical layout is shown in Figure 2.11, the 'white water' being the recycled effluent saturated with air. There are some rectangular tank designs with a raised beach at one end and a travelling scraper. It is important here that sludge dripping from the returning scraper does not knock down floating sludge.

Sludge dry matters of 4%+ should be expected from DAF plant, and 10% is achievable with oily sludges. Because some solids will still sink by this process and some separation of different materials can therefore be achieved, flotation lends itself readily to recovery of valuable raw materials, and a clean, consistent floating product. In some instances, it will be possible to incorporate this back into the production line directly.

In sewage treatment, DAF has been used successfully to separate thickened activated sludge, the process initially encouraged by allowing natural denitrification and some prethickening to take place under quiescent conditions; 8% DM has been achieved, remarkable for activated sludge, but the system has found no long-term favour on sewage works because of inconsistent performance and difficulty in inducing reliable denitrification. Conversely, it has wide application in industry, and is used very successfully to treat dairy water high in fat and protein; 75% removals of BOD load are routine performance levels, where DAF can be used in place of high-rate roughing filters to reduce effluent strength to that of crude sewage (250 mg/l BOD, 300 mg/l suspended solids), prior to conventional secondary biotreatment treatment. If high rate biofilters can be avoided and substituted with DAF, so too can the odour problems associated with them — a particular benefit on cramped sites.

Figure 2.12 shows the floating product on a DAF plant treating a dairy waste, and Figure 2.13 the chemical mixer and pH correction tank upstream. This particular site is described more fully in Chapter 8 as a case history. In some instances where DAF is used to treat dairy or food industry waste, the effluent, after passage through a clarifier, may be acceptable for discharge without further biological treatment, and certainly acceptable as a trade

46 Sewage and Industrial Effluent Treatment

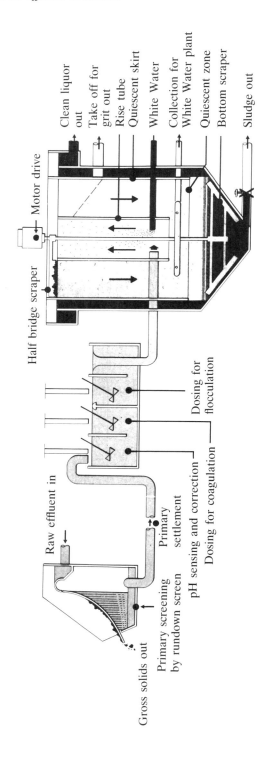

Figure 2.11 Layout of DAF plant.

Figure 2.12 DAF and associated equipment – dairy effluent plant.

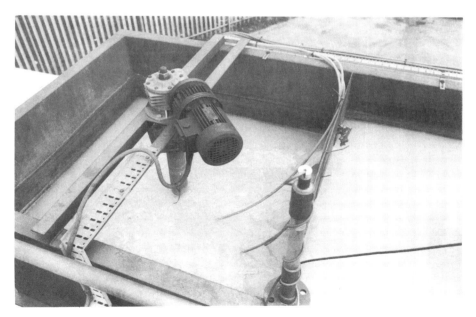

Figure 2.13 Chemical mixing and pH correction for DAF plant.

48 *Sewage and Industrial Effluent Treatment*

effluent to the foul sewer. Other advantages of DAF here include the prospect of recovering the high fat//protein sludge, where it may have value as an animal feed and so turn an expensive tankering and disposal operation into a potential income source.

There are numerous other potential applications for using DAF in industrial effluent treatment. Canning fruit and vegetables produces much organic suspended solid material amenable to removal by DAF. Inorganic fines produced during the polymerisation of synthetic rubber, polyethylene, and carbon powder, colloidal metal and metal hydroxides are other applications.

Coal and pyrites fines are separated from slate by injecting carbon dioxide gas, and flotation can be used in place of settlement basins in coal washing plant.

The iron and steel industry uses DAF to capture scale and oil, and the oil industry to separate emulsions and water contaminated by oil at 50–400 mg/l levels. The latter effluent can be treated in a two-stage package plant consisting of two upward flow tanks, having first been injected with air in a mixing chamber pressurised to 6 bar. Polyelectrolyte and a flocculant are

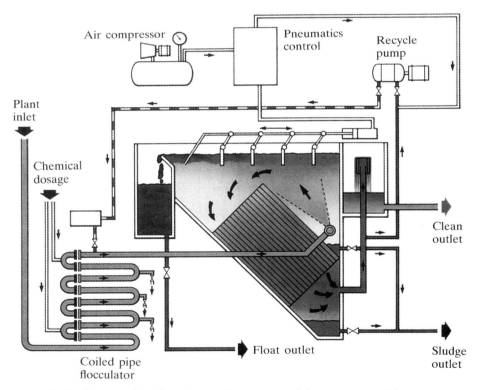

Figure 2.14 Cross-section through an oil–water emulsion separator. (Courtesy: Environmental Engineering Ltd.)

added and the pH adjusted. The flow pattern in this unit is shown in Figure 2.14. This particular unit will separate a wide range of emulsions and solids-laden effluents. The suppliers are Environmental Engineering Ltd.

The DAF technique is commonly applied as an aid to clarification and colour removal during potable water treatment, which may be continuously or seasonally required as the humic acid content increases, polyelectrolytes and aluminium sulphate being effective treatment chemicals.

2.9 Common problems associated with settlement, clarifiers and DAF

2.9.1 Settlement

(1) The tank(s) are hydraulically overloaded at peak flows. This is normally seen as a sudden burst of rising sludge. Check the UFR at peak flow times; flow balancing upstream may be necessary.

(2) Generally poor settlement and a consistently poor solids removal of less than 50% can be caused by several factors.

    (a) The solids may not be amenable to settlement except under completely still conditions. If a fill and draw operating method cannot be implemented, the addition of a coagulant/flocculant to increase particle size may effect better settlement. Fines can often be successfully trapped by a wedge wire clarifier or fine filter cloth placed across the tank, usually 0.1–0.3 m below top water level. Such filters should be checked as they tend to 'blind' quickly.

    (b) Scum boards fitted to contain floating material on the tank surface should not be more than 0.2 m deep. The author has seen a number of examples in the metal treatment industry where the scum board was within 0.3 m of the tank floor, causing settling solids to be swept under and up into the effluent launder. An immediate cure was effected by cutting the board to a quarter of its original depth!

    (c) The tank may be hydraulically overloaded and the UFR too high. Solutions are to increase settling volume or balance the flow, either involves additional tank(s). The insertion of lamella plate separators in the tank reviewed in Chapter 6, will also increase settling area. This will be a cheaper option than building new tanks, quicker to implement, and provide some flexibility if effluent solids fluctuate seasonally.

(3) The solids loading is too high. In this case, it is prudent to examine first where the solids are coming from and whether levels can be reduced by work practice changes. The desludging frequency or sludge withdrawal rate should also be examined (see Figure 2.2). In many cases, the problem simply goes away by regular, frequent desludging, rather than by waiting until solids are visibly coming over the weirs!

(4) Warm effluents may stratify into layers, causing variable settling efficiency,

and often dependent on ambient temperatures. In this case, some form of in-line mixing or baffles placed upstream will often improve settlement from different effluent streams. Mechanical mixing in a separate tank may be necessary. Poor settling caused by turbulence indicates a revision of the inlet arrangement to the tank(s). In rectangular tanks, an inlet channel with a continuous baffle which causes the flow to spread out over the tank area is required. Inlets arranged around the tank(s) are another option.

Circular tanks with a central inlet pipe are ideally fitted with a diffuser cylinder surrounded by a stilling box of adequate depth. Figure 2.15 shows this arrangement.

(5) Pumping wastewaters with organic solids between the biotreatment and settlement stages can often cause shearing of delicate sludge flocs and the generation of fines which will not readily settle and impart a cloudiness to the effluent. Siting the settlement tanks higher than the biotreatment outfall is not good design and should be avoided. The author has seen a number of small treatment works where this has been attempted merely

Figure 2.15 Central inlet baffle and stilling box – sedimentation tank.

because it was more convenient to construct prefabricated tanks at ground level than to excavate. The resulting high solids effluent did not improve with passage through a microstrainer because the particles were mostly smaller than the strainer mesh size.

2.9.2 Clarifiers

(1) Problems with these are usually confined to sludge blanket loss, caused by sudden flow variation. Flow balancing upstream is the usual solution, but avoiding lengthy retention times likely to cause organic sludges to destabilise. Some form of chemical dosing is commonly used with clarifiers.
(2) If wide flow variations are unavoidable, the response time of the sludge blanket detection system, the depth of maintained blanket and the sludge withdrawal rate should all be examined. Equipment reliability in this area of operation is vital.

2.9.3 Dissolved air flotation

(1) In addition to examining the hydraulic and solids loading rate, a check should be made of the physical state of the air nozzles and compressor air pressures if solids removal is poor.
(2) The scraper mechanism can 'knock down' surface solids if rough in action or material falls from it. Cyclic cleaning of the blade or brush will prevent this problem. Rubber blades often degenerate rapidly in industrial effluents and adopt a rigid 'curl'; stiff polyproplene bristles are usually more flexible for longer and the flicking action over the discharge trough self-cleaning.
(3) As both scrapers and air supply equipment must work continuously in DAF, security of electricity supply and mechanical reliability are fundamentally important for consistent performance.

2.10 Conclusions

There are many sources of reference in the literature to the techniques described in this chapter. UK readers are referred to the Journals of the Chartered Institution of Water and Environmental Management (CIWEM), whilst other countries will have their equivalent learned societies with interests in water pollution and treatment.

Although the solids separation techniques described will often reduce effluent strength by 50%−90%, most organic wastewaters will need a biological oxidation stage before discharge, and these are described in Chapter 4.

2.11 Summary table of typical operating parameters and dimensions

2.11.1 Settlement tanks

Suspended solids percentage removal	40%–75%
BOD percentage removal	20%–80%
Bacterial removal (primary tanks, sewage treatment)	10%
Upward flow rate (UFR)	$0.5\%-1.5\,m^3/m^2/h$
Retention time for 50% solids removal	2–8 h
Maximum solids loading (MSL)	$7.8\,kg/m^2/h$
Stirred specific sludge volume (SSV)	80–140 ml/g at 3.5 g/l SS
Rectangular tanks ratio of width to length	1:2–1:4
Circular tanks ratio of diameter to depth	3:1–8:1

2.11.2 Upward flow clarifiers

Upward flow rate (UFR)	$0.5-1.0\,m^3/m^2/h$
Ratio of diameter to depth	0.5–0.8
Typical sludge blanket thickness	0.3–0.9 m
Sludge blanket rise rate	3–5 m/h
Ratio of minimum flow (to just maintain a sludge blanket) to the maximum (which causes it to rise and expand suddenly)	3:1

2.11.3 Dissolved air flotation

Satisfactory performance primarily depends on the ratio of the volume of air (A) to the mass of solids (S). A/S is typically 0.005–0.006.

BOD percentage removal	75%+
Upward flow rate (UFR)	$3-7\,m^3/m^2/h$
(Sludge thickening is often achieved successfully at only $1-2\,m^3/m^2/h$).	
Solids loading rate	$10\,kg/m^2/h$
Recycle ratio	15%–100%

2.12 References

Jones, K., Smith, D.E. & Thomas, C. (1991) The application of physicochemical treatment to an overloaded sewage works. *Journal of the Institution of Water and Environmental Management* **5**(1).

WRC (1975) *Technical Report TR11*. Water Research Centre Publications, Medmenham, Buckinghamshire.

2.13 Companies and other organisations

Environmental Engineering Ltd, Little London, Spalding, Lincolnshire PE11 2UE.
Chartered Institution of Water and Environmental Management, 15 John Street, London WC1N 2EB.

Chapter 3
Metal Removal Methods

3.1 Introduction

Most methods of removing or reducing metal levels in effluents from industrial metal finishing processes are physical or employ physical chemistry techniques, precipitation and electrolysis being examples. There are three primary reasons for such treatment.

(1) *Economic*. Most plating solutions represent a considerable raw material investment and their slow loss, except as product on the workpiece, is financial folly. Recovery of metals for subsequent re-use, while it may involve a third party who takes sludges or spent ion-exchange resins away, should be a principal activity in metal treatment.
(2) *Environmental*. Most of the common metals used in the metal treatment industry – chromium, nickel, zinc, copper, cadmium – are very toxic to the aquatic environment and sewage treatment processes. Cyanide is extremely toxic both in solution and as a gas liberated under acidic conditions.
(3) *Regulatory*. With very few exceptions, the discharge from any metal finishing process will be classed as a trade effluent to the foul sewer and concentrations measured in hundreds of grammes per litre during, say, plating, must be reduced to less than 10 mg/l in many cases, i.e. reduced by factors of between 10^4 and 10^5. In the few cases where discharge is to a watercourse, the metal content is likely to have to be reduced to negligible levels if considerable damage to the ecology of the stream or river is to be avoided.

There is some contemporary interest in biological systems for metal removal, normally by propagating metal-tolerant plants grown in the effluent from a conventional biological treatment plant to give a final clean-up. There is also interest in the development of metal-tolerant fungi and bacterial cultures to act as bio-extractants and this is reviewed in Section 3.6. Perhaps the most significant developments recently involve closed loop systems where total recovery of metals by electrolysis or ion-exchange and recycling of all water occurs. These methods are gaining ground in Europe because the trend to tighter effluent standards has highlighted the limitations of conventional precipitation methods using lime or sodium hydroxide.

Not all effluents associated with metal finishing contain metals in any form; instead simple pH adjustment is required of acid or alkaline solutions and this aspect is reviewed hereunder.

3.2 An overview of effluent treatment options from metal finishing processes

Metal finishing is a general term covering a range of activities including plating, anodising, polishing, cleaning, pickling and scale/oxide removal. Metals may be added to surfaces or removed by abrasion and chemical cleaning – either way solutions and liquids involved with these processes will contain metals or become contaminated by them. Any resulting effluent will similarly contain metals in dispersion or solution and normally require some form of treatment before discharge is permitted.

Plating solutions are either alkaline or acidic and require pH correction within the range 8–9.5 to precipitate the metals as the hydroxide. Initially it is often easier to treat acid and alkaline plating solutions separately. Cyanide effluents should also be kept separate and the cyanide oxidised to cyanate before any other treatment or mixing takes place. Chromic acid and chromate rinse waters must be reduced to the trivalent state before the chromium can be precipitated out and careful pH control exercised.

Cadmium is very toxic and many platers have ceased handling it. The main use remains in specifications for military equipment where the alternative, nickel, causes unacceptable embrittlement. Cadmium also provides superior salt-water corrosion resistance. Currently the regulatory bodies impose limits of less than 0.5 mg/l which preclude simple hydroxide precipitation and in some cases have excluded the metal. Ion exchange, electrolytic capture or complete closed-loop plating are thus the few options for this element.

The common mineral acids – sulphuric, nitric, hydrochloric and phosphoric – are used at varying concentrations for anodising, cleaning, etching, pickling and de-scaling metal surfaces. Alkaline cleaners in common use are either based on sodium hydroxide or sodium carbonate and often contain phosphates for their water softening action, silicates and surfactants for wetting and emulsifying and chelating agents like EDTA. Detergents are often included in proprietary mixtures.

Apart from recovery and cyclic re-use, industry employing strong acids and alkalis must control the pH of the effluent from such processes, usually in the range 6–10, before it enters the sewer. This is necessary primarily to avoid damage to the fabric of the local and public sewers including the metalwork at access points.

Often effluent flows can be mixed to achieve some neutralisation, but the presence of metals either from the original reagents or resulting from corrosion and chemical attack of pipes and tanks sometimes dictates a metal removal stage even although the manufacturing process itself does not use metals or their solutions. Zinc and copper are particular examples.

Metal de-greasing and cleaning with solvents is rapidly diminishing. Tetrachloroethylene and trichloroethylene were once widely used, but the difficulties of safe handling and capacity for atmospheric environmental damage have resulted in a ban on their use in the near future in many Western countries. All the chlorinated solvents are very toxic to sewage treatment processes, particularly sludge digestion. They are specifically banned from most trade effluent Consents and consequently have always been recycled to the original supplier for cleaning. They must never be discharged to sewer or watercourse – there are very elegant and sensitive analytical methods for their detection at 10^{-15} levels.

Mechanical abrasion and cleaning processes, including the output from iron foundries, will often contain metal particles in suspension. Simple settlement, with a flocculating polymer if found beneficial, will often suffice; the sludge may have a resale value. Coarse filtration or lagooning are alternatives. Iron, while not in itself toxic to sewage treatment, can induce septicity under acidic conditions and the sludge can readily block small diameter drainage systems.

3.3 Effluent toxicity and discharge limits

Most solutions and chemicals connected with metal treatment are complex mixtures with a number of components of wide-ranging toxicity. The principal interests of the regulatory authority relate to preserving the sewer fabric, safety of those at pumping stations or working within the sewerage system and maintaining efficient sewage treatment operations.

As a general guide, all the metals associated with plating are toxic and levels of less than 10 mg/l are set individually in Consents. Cadmium, a 'Red List' substance in the UK whose limits are set by HMIP and not the Water PLC, is now prohibited in some discharges or set at the very low limits of 0.02–0.05 mg/l.

In practical terms, any form of simple hydroxide precipitation is thus precluded by these limits, closed circuit ion-exchange being a favoured option.

The UK Ministry of Defence remains one of the few specifiers of cadmium plating for air frames (the nickel alternative causes embrittlement). Domestic applications of cadmium have been further restricted in the UK by the Environmental Protection (Controls on Injurious Substances) (No.2) Regulations 1993 which enact EC Directive 91/338/EEC. Apart from the toxicity of cadmium to sewage treatment works operation and particularly sludge digestion, it is limited to 4 mg/kg dry matter (DM) in any sludge applied to farmland. Thus the level that could be acceptable in crude sewage, allowing for concentration in the sludge, is unlikely to exceed 0.002 mg/l.

Alkaline cleaners containing phosphates and detergents are much less problematic although phosphating mixtures will contain other metals like zinc and copper which will be severely restricted. In the long term, phosphate limits may be imposed in areas where the sewage works is significantly adding to the nutrient load of the receiving waters. Other anions, such as chlorate and nitrate found in phosphating solutions and fluorides, are unlikely to be restricted by most Consents allowing discharge to sewer. Solvent-based cleaners are virtually banned from wastewater discharges in most countries, or soon will be.

Mixtures with organic chemicals often have specific limits for these set; allylthiourea is an example of a material that is very inhibiting to nitrification during sewage treatment. Others may not readily break down during normal sewage treatment; all will contribute to the Chemical Oxygen Demand (COD) 'strength' factor in the trade effluent charging formula.

Aluminium from anodising is not normally a problem and the chemical is often added to sewage sludge for conditioning purposes or to assist sedimentation. In fact, it can be quite beneficial in small doses although the effluent inspector is unlikely to reveal this!

Iron from pickling and de-scaling is of similar low toxicity although it can trigger septicity in long rising mains and sulphide production. High sulphate levels can induce the same problem. With both these elements and from metal cleaning generally, the level of solids, metallic or otherwise, is likely to be more of a problem and less than 500 mg/l suspended solids (SS) limits are often imposed.

Table 3.1 lists metals and the levels at which they are likely to cause toxic effects to various stages of sewage treatment. These manifest themselves in the biological oxidation stages as a loss of nitrification, sloughing of solids from biofilters and poor settling or bulking sludges. Extreme cases result in a complete loss of oxidation, a rapidly dying biomass and considerable odour problems. The same effects will be seen on the flora of a stream; the toxicity to fish (rainbow trout is the usual standard) of a number of metals is listed in Table 3.2.

Although metal settlement in the primary sedimentation stages of sewage treatment varies depending on the element and the physics of the tanks, an average 75% of incoming metals in the crude sewage collect in the raw sludge. In the subsequent biological stages, the majority of the remaining 25% of (mostly soluble) metals are adsorbed by the biomass, some trace essential elements, e.g. selenium, being retained for long periods by activated sludges. Ultimately, therefore, most of the metals entering a works will end up in the sludges and only about 10% will pass through to the effluent discharge.

Sludges high in toxic metals present operational and ultimately disposal problems. If incinerated, the ash will contain a high concentration of metals, four or five times the level originally in the sludge, and this might restrict disposal options. However, at normal ground pH values, and assuming no

Table 3.1 Metal levels toxic to sewage treatment processes.

(a) *Anaerobic digestion*

Metal	Concentration range (mg/kg DM)
Zinc	30–140
Nickel	7–50
Cadmium	200–800
Copper	500–3000

(b) *Crude sewage levels likely to inhibit biotreatment stages and digestion*

Metal	Concentration (mg/l)
Zinc	2.2
Nickel	2.0
Lead	7.0
Cadmium	3.8
Copper	3.2

Notes to Part (a):
1. Values are for a 20% reduction in gas yield.
2. Effect varies with solubility and therefore pH and sulphide concentration.
3. For mixtures, the effects are additive on an equivalent weight basis or meq (milligram equivalent weight)/kg DM. Thus:

$$K(\text{meq/kg}) = \frac{(Zn)/32.7 + (Ni)/29.4 + (Pb)/103.6 + (Cd)/56.2 + (Cu)/47.4}{\text{Sludge solids concentration in kg/l}}$$

If K is 400 meq+, there is a 50% chance, of digestor failure.
If K is 800 meq+, there is a 90% chance of digestor failure.
If K is <160 meq, there is a 90% chance of digestion being unaffected.

Note to Part (b):
Nitrification is impaired above 1 mg/l concentration for individual metals.

Values are courtesy of CIWEM.

Table 3.2 Metal levels toxic to trout.

Metal	Rainbow trout 48 hour LC50 (mg/l)
Zinc	1.4
Nickel	43.0
Lead	1.0
Cadmium	0.9
Copper	0.16

1. The 48 hour LC50 is the concentration that kills 50% of the fish in 48 hours.
2. The toxicity is greater in soft water and at low DO concentrations. The above values are for water 50% air saturated and with a hardness of 100 mg/l as $CaCO_3$.
Values are courtesy of CIWEM.

elution with acidic leachate, ash from this source is inert and the metals quite tightly 'bound'. Each disposal site will need monitoring and the problem will be compounded should the EU classify this ash material as toxic in the future.

There is the possibility of metal recovery from sewage sludge ash; some sewage sludge ashes from industrial areas contain levels equivalent to mined ore at 2%+, but to date, little work has been done on recovery techniques. Some volatile metals, e.g. cadmium and lead, will be trapped by subsequent scrubbing or dust collection in the incinerator and re-enter the treatment works in return liquors.

Anaerobic sludge digestion is inhibited by metals and lower gas yields cause operational difficulties in keeping the digestor up to temperature.

Sludges transported to land for their fertiliser value and containing toxic metals are limited in application rate and by the time of year and crop type. Fields previously sludged may be unable to accept more if sludge metal levels are high, and there is increasing concern in farming circles about the long-term build-up of toxic metals on farmland. Sludge disposal costs then rise.

Thus, four factors conspire in the setting of tight metal limits by operating and regulatory bodies: a loss of works treatment efficiency, operating difficulties, restricted sludge disposal routes and general toxicity in the aquatic environment. The UK Water PLCs are noticeably tightening current metal limits in trade effluent Consents. Typical limits for metal-containing trade effluents discharging to the sewer in the UK are set out in Table 3.3. There are regional variations depending on dilution, past history and type of receiving works but the general trend for limits is ever downwards.

Table 3.3 Typical UK trade effluent Consent limits.

pH range	6–11
Settleable solids	300 mg/l
Total cyanide	5 mg/l
Sulphate	1500 mg/l
Total sulphide	1 mg/l
Available chlorine	50 mg/l
Available sulphur dioxide	5 mg/l
Free ammonia	100 mg/l
Grease and oil	500 mg/l
Chromium	10 mg/l
Zinc	10 mg/l
Nickel	3 mg/l
Copper	5 mg/l
Lead	10 mg/l
Silver	1 mg/l
Cadmium	subject to EC Directive 76/464/EEC and HMIP approval (typically <0.1 mg/l)
Total iron	100 mg/l

All halogenated hydrocarbons and petroleum spirit prohibited.
Temperature not to exceed 43°C.

3.4 Treatment methods for metal-containing effluents

3.4.1 General aspects

Because there are discharge facilities and the effluent must contain substantially lower levels of metals than those of solutions in use during manufacture, some form of treatment will certainly be required. Diluting the effluent with a running hose to achieve the standards set is not one such form of treatment, although the author has seen it practised. Apart from being an immoral waste of water, water consumption costs will in time prove much greater than even the more sophisticated treatment plant. This equally applies to the hose discreetly flooding the floor continuously; if required to wash down items, it should be fitted with a trigger valve to permit use only when needed.

The possibility of combining waste flows using one to treat another should be examined first. Acids and alkalis are obvious examples for even if there is a shortfall of one reagent, some savings can be made. This method is most appropriate where the metal content is either low or not the greatest treatment problem and cyanides are not involved. But many effluents will demand segregation for the most economic and effective treatment; likewise, the production lines are often better separated – or thought should be given before mixing.

Housekeeping and layout are important. Tanks between which work is being transferred should be grouped together so that spillage is minimal. Adequate draining time should be allowed during transfer and a static dragout tank employed as the first catch-all after plating operations as this will allow recycling of valuable plating solutions. Gold and silver platers rarely spill anything on the floor. By contrast, some small firms plating nickel, chromium and zinc drop concentrated solutions of these metals whose combined mass could be calculated as hundreds of kilograms per annum. The author has visited a number of such companies over the years.

Choosing a treatment plant will depend on the effluent discharge quality dictated by the regulatory body, daily volumes, discharge rate and on-site expertise. Small discharges less than $20 \, m^3/day$ are best dealt with on a batch basis and manually treated. Two tanks will be required, one for collection and another for treatment.

Continuous flow plant may need buffering or flow balancing and there may need to be some pretreatment of individual discharges. There must be adequate mixing and retention time for reaction and to measure or sample the product: electrodes take finite times to react and dose chemicals as required and some plant are volumetrically too small for automated measurement systems to react and control properly before the effluent is down the drain.

All the control systems on an effluent plant must be regularly serviced. Electrodes must be checked and recalibrated – daily if past experience calls for this. Valves and dosing pumps must be checked for full and correct operation along with the voltage they are running at. When all the baths are switched on, the voltage drop in some buildings can be surprising, and 20% is not unknown. Solenoid valves then 'chatter' and fail to deliver or close properly and pump output is reduced or erratic.

In most circumstances, the cost of disposing of the metal-laden sludges will be much greater than that of the reagents to treat the effluent. Where a choice of treatment chemicals is available, always choose those which offer the minimum sludge volume of the greatest density.

3.4.2 Cyanide-free and chromium-free effluents containing zinc, nickel, copper and cadmium

These effluents are treated by adding milk of lime, sodium hydroxide or sodium carbonate solution to achieve a pH in the range 8.5−9.5. Metals will be precipitated as the hydroxide − a light, fluffy sludge which will settle in quiescent conditions and normally without the aid of polyelectrolytes. The optimum pH value for individual metal precipitation as the hydroxide is shown in Figure 3.1. The lime consists of powdered hydrated (slaked) lime in water and to keep it in suspension, constant stirring of the make-up tank is necessary. This is the cheapest material but will generate more sludge for disposal than the other two chemicals, although the sludge is denser and tends to settle more quickly. Hydroxide sludges generally dewater well and a 40%+

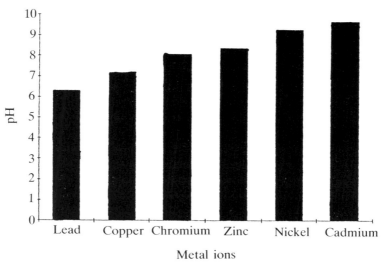

Figure 3.1 Bar graph of optimum pH values for metal ion precipitation as hydroxides.

dry cake should be obtained by most of the mechanical dewatering techniques described in Chapter 5.

In-line mixing and stirring facilities should be adequately sized to allow pH adjustment to a stable value before settlement. If ferrous iron is present, an air sparge after settlement followed by final clarification will oxidise this to the ferric state; combined with other metals in the hydroxide form, further settlement will then take place. The effluent should contain less than 5 mg/l of any metal, and less than 2 mg/l can be achieved. Poor settlement and solids carryover are the most likely causes of higher levels and, when these occur, the design of the settlement tank(s) should be examined or a flocculating aid considered.

3.4.3 Chromium-containing effluents

Chromium is present as chromic acid or chromates and must first be reduced to the trivalent state. Two reducing agents are commonly used: sulphurous acid and ferrous sulphate.

Sulphurous acid is added either as sulphur dioxide gas or sodium metabisulphite solution. The pH of the effluent must first be reduced to $2-2.5$; it is sometimes possible to make use of other acidic effluents for this purpose, or sulphuric acid is used.

Bisulphite is easier and safer to store and handle, the theoretical dose being 3 parts bisulphite to 1 hexavalent chromium; 0.6 part sulphuric acid would be needed to reduce the pH to 2.5 in the presence of 1 part chromic acid. If the incoming chrome swill is at pH 7, 700 ml of a 20% sulphuric acid solution will reduce the pH to 2.5. Using sulphur dioxide gas, 0.8 part will neutralise 1 part chromic acid. Ferrous sulphate (copperas) requires 16 parts of the hydrated form to 1 part of hexavalent chromium and produces much more sludge on neutralisation. This disadvantage must be balanced against the lower cost of this reagent.

Following chrome reduction, the effluent is mixed with lime solution, sodium carbonate or sodium hydroxide to raise the pH to $8-8.5$ and precipitate the trivalent chromium. This can be combined with other metal precipitation. By any of the above methods the effluent should contain less than 2 mg/l chromium.

3.4.4 Cyanide-containing effluents

These should be treated before any subsequent dilution, to maximise removal rate and the benefit of later dilution. Where discharge limits are less than 10 mg/l, cyanides should be separated from liquors containing copper and nickel as stable complexes soluble in water may form and are not easily decomposed.

Acid and cyanide streams must always be separated; hydrogen cyanide gas liberated by acidic conditions is extremely toxic and cyanide oxidation to cyanate is easily reversed by acids, the hydrocyanic acid formed being a lachrymatory gas. The intermediate cyanogen chloride is also toxic and all reactions should be carried out in closed or force-ventilated tanks.

Two processes are used to treat cyanide effluent and these are now briefly discussed. The older method uses ferrous sulphate and lime. Both chemicals are cheap but only achieve partial reduction of cyanide and considerable sludge volumes are produced. The reaction is best carried out in the pH range 7.5–9; this method is rarely used in Western Europe where few trade effluent Consents permit total cyanide discharge levels of more than 5–10 mg/l.

By far the more common method today is that of adding chlorine gas or sodium hypochlorite solution; large plating shops use chlorine gas but the smaller firm will find that handling and safety considerations dictate the latter. Chlorination is carried out under alkaline conditions and cyanogen chloride is initially formed, the intermediate oxidation product. Sodium carbonate or hydroxide is used to adjust the pH and, if kept above 10, rapid hydrolysis yields the cyanate which converts slowly to ammonium carbonate; cyanate is about 200 times less toxic than cyanide. Between pH 8.5 and 10 and in the presence of excess chlorine, cyanogen chloride and cyanate decompose to nitrogen gas and carbonate. Approximately 2.7 parts chlorine gas and 3 parts sodium hydroxide are required to destroy 1 part cyanide.

If hypochlorite solution is used, the most economical operation is carried out at pH 10 or above. Roughly 2.5 parts sodium hypochlorite are required to 1 part sodium cyanide, the hypochlorite being available as a 15% solution. If a batch process is used, 15 minutes' reaction time with stirring is recommended. Starch iodide paper will turn blue in the presence of free chlorine and is a simple method of checking. The reaction proceeds thus:

$$CN^- + ClO + H_2O \rightarrow CNCl + OH^- \rightarrow CNO^- + Cl^- + H_2O$$

Excess chlorine can be removed by sodium thiosulphate; some Consents may require this and 2 parts of the crystalline form will neutralise 1 part excess chlorine. If sulphur dioxide is available from chromate reduction, this will also remove excess chlorine.

Where zinc is present, the effluent will require a reduction in pH to 8–9.5 followed by settlement, as at very high pH, the soluble zincate is formed.

These methods are suitable for cyanide effluents up to 500 mg/l and very strong solutions/spillages will need to be either bled very slowly into a continuous treatment plant or disposed of separately. The effluent should not contain detectable cyanide levels if dosing is correct and adequate reaction time given.

3.4.5 Cadmium-containing effluents

Cadmium cyanide and its complexes is treated initially as in 3.4.4 above. The element will precipitate out as the light, fluffy hydroxide in the pH range 8–10, in common with other plating metals, as in 3.4.2 above. The effluent is likely to contain less than 1 mg/l if the settlement stage is quiescent and regularly desludged. Consent limits of 0.02 mg/l or exclusion of cadmium altogether from the discharge dictate other methods including ion exchange and electrolytic recovery outlined later.

3.4.6 Anodising

Both sulphuric and chromic acid are mainly used, the former in 10–50% solutions. The chromic acid can be treated in the way described in 3.4.3. Rinses containing sulphuric acid will require neutralising to pH 8–10 and quiescent settlement is required to allow the light flocs of aluminium hydroxide to settle. Sodium hydroxide is the easiest treatment chemical to use and can be dosed as a dilute solution via a metering pump controlled by a pH electrode. As a guide, 165 ml of a 20% solution of NaOH will raise the pH of 1 m^3 of effluent from 7 to 11.

Upstream flow balancing is recommended and a regular de-sludging programme. Disposal of spent anodising baths must be carried out very slowly. The dyes used in anodising are not normally a problem in the discharge and often adhere to the hydroxide flocs.

Anodising tin, silver, copper or zinc will require stricter attention to settlement technique to conform to metal discharge limits. Aluminium and magnesium are often not listed in Consents because of their low toxicity and ubiquitous nature but in practice will be limited by the suspended solids limit, typically 300–500 mg/l.

3.4.7 Pickling solutions

These are normally composed of sulphuric or hydrochloric acid; phosphoric acid is widely used to remove rust from steel. Apart from maximising recovery for re-use, pH correction and settlement are the two primary requirements in the largest tanks that can be accommodated. This will enable thorough mixing of reagents, adequate reaction time and the use of any surplus alkaline reagents.

Remember that pH is the log of the H+ ion concentration and, being a logarithmic scale, the quantity of alkali needed to raise the pH from, say, 3 to 4 is only one-tenth of that to raise it from 2 to 3 while between pH 6 and 7, only 1/10 000 of the volume is needed. This makes automatic control of pH around pH 7 with a single electrode and on/off dosing valve difficult.

Pickling is designed to remove considerable quantities of scale and oxide layers, and these layers must be removed regularly by desludging; 300–500 mg/l solids is a frequent Consent standard. Trade effluent Consents often also limit sulphate to 500 mg/l and some dilution or mixing with other discharges may be necessary.

3.4.8 Metallic particles

From cleaning and machining activities, oil is a frequent contaminant of metal particles and its recovery essential. Some form of oil separator or flotation tank is then required as outlined in Chapter 1 prior to gravity settlement of the heavier particles. In view of the density of most metal particles, initial agitation of the effluent may be necessary, which will also assist oil/water/metal separation. Finely divided metals can also be recovered by dissolved air flotation in the same manner as fine particles of coal or slate, using frothing agents.

It is essential that any solubilisation of particles is avoided and that acid/alkaline discharges are kept quite separate from particulates until neutralised and the bulk of the particle settled.

3.4.9 Photo-etching wastewater

The main components are: 1000−2000 mg/l ferric chloride, about 200 mg/l ferrous chloride and 400 mg/l cupric chloride; 50 mg/l of nickel, chromium and zinc chloride may also be present. The effluent can thus be treated satisfactorily by pH adjustment to 8.5−9.5, as outlined in Section 3.4.2, and settlement of the hydroxide sludge. Polyelectrolytes or lime may prove useful to improve sludge settlement characteristics. The low toxicity of the main component, iron and low levels of the other metals will simplify disposal options as the material will have no value. Such sludges with an iron and lime content often dewater well and a small plate press will produce a cake of over 40% DM which may be taken away by skip, a much cheaper and less frequent option than liquid tankered disposal.

3.4.10 Ion exchange

This is primarily used to recover valuable or rare metals and widely applied in the photographic and electronics industries, in which gold and silver are used. Another principal application is to trap and prevent the discharge of prohibited elements, particularly cadmium and other more unusual components like radioactive isotopes.

In all cases, the materials are trapped on a resin bed by substituting the toxic metal with a cation of much lower or insignificant toxicity; the metal is retained and the substitute eluted. Anion resin columns similarly remove relevant anions − sulphate, cyanate and silicates.

When the resin bed is exhausted, it is either regenerated *in situ* (in which case the eluate produced containing the concentrated toxic metal, is recovered, recycled or sent for processing), or the whole resin column is exchanged for a fresh one and the contents of the original recovered by the supplier. The latter is the more common method, being more convenient for many operators.

The supplier pays to take away the resin in most cases, the amount paid depending on the value of the trapped material determined by analysis. Extraction, transport and supply costs and a modest hire charge for the column are deducted.

A specific example is the use of ion exchange where the plating shop has no access to a sewer and a closed-loop recirculation system is required. Mains water supplies the rinse tanks and the effluent passes to the ion exchange plant. Cyanide is oxidised to cyanate and hexavalent chrome to the trivalent state first. An in-line filter to remove large particulates, UV steriliser to kill bacteria and algae and an activated carbon filter to remove organic matter are often placed upstream of the main resin columns.

A cation column then removes heavy metals and is regenerated by hydrochloric acid; the anion column removes sulphates, cyanates, carbonates and silica and is regenerated by sodium hydroxide. Regeneration is controlled automatically by a conductivity probe, and activates when the column effluent exceeds 5–25 µS/cm (or whatever quality standard the plating shop requires). Used regeneration chemicals are stored and disposed off-site by a hazardous waste contractor.

Mains water is only used for top-up due to evaporative losses. As a bonus, the recycled water is softened, lime and carbonate scale absent and less work rejected by quality control.

3.4.11 Electrolytic methods for metal recovery

These are used to electro win metals from solutions which can range in concentration from 1 mg/l to 200 g/l; in practical applications, most operate in the 50–250 mg/l range in closed loops. The metal is plated onto the cathode in a pure form which is periodically removed.

Most available commercial cells have two common features: a high electrode area is achieved by the use of gauze or expanded metal and fluidised glass beads distribute flow and ensure efficient mass-transfer. Often they are inserted in recycle streams to prevent metal levels building up; they vary in removal efficiency.

One commercially available example, the Chemelec cell available from BEWT (Water Engineers) Ltd, has the standard dimensions 0.5 m × 0.6 m × 0.7 m, and is equipped with six stainless steel sheet cathodes (total area 3.3 m^2) and seven platinised titanium mesh anodes. The system is modular and five sizes are available. Flow passes upwards; the layout in Figure 3.2 includes a reservoir tank and other models allow for pH control and carbon filtration of the effluent which discharges from the top weir.

Ideally, the metal solution should contain only one metal; the plated cathode can then be removed and used as the anode in a plating tank or sold as electrolytic grade material. Metal recovery better than 99% can be achieved and the cell will recover from metal solutions as low as 20 mg/l. The rate

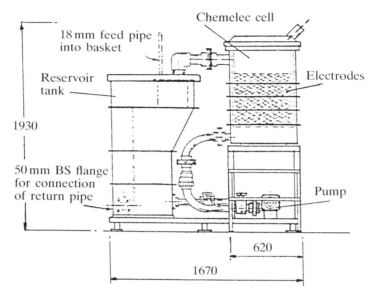

Figure 3.2 Cross-section of the Chemelec cell. (All dimensions are in mm.)

of metal recovery depends on operating conditions; Table 3.4 provides a guide to the standard model. Power consumption is about 6 kWh; total power consumption including pumping is quoted at 7.5−20 kWh/kg of metal recovered.

These units are commonly used to control the metal level in the static rinse immediately after the plating bath; Figure 3.3 shows a typical layout. The cell reduces 150−200 mg/l metal inputs to about 50 mg/l. Subsequent rinses are then less loaded with toxic metals carried forward and the effluent complies with most Consent limits.

Other manufacturers of electrolytic cells recommend their application in similar situations; a few offer configurations that will reduce metal effluent levels to 5 mg/l or less, permitting direct discharge, but these are in the minority.

Table 3.4 Metal removal rates by the Chemelec cell.

Metal	Maintained concentration (mg/l)	Rate of removal (g/hr) − standard model
Silver	50	125−250
Gold	20	40−60
Copper	500	250−375
Nickel	500	185−250
Zinc	500	125−185
Cadmium	50	60−95

68 *Sewage and Industrial Effluent Treatment*

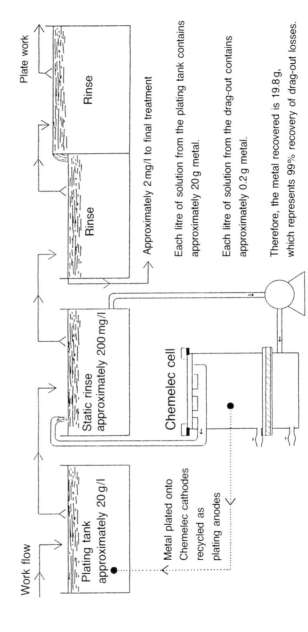

Figure 3.3 Plating shop layout with Chemelec cell.

Although electrolytic cells have a high capital cost and energy running costs, their main advantages include the recovery of valuable metals, non-generation of toxic hydroxide sludges and associated disposal costs, no chemicals and the anodic destruction of complexing agents and cyanides.

3.5 Plating shop layout and housekeeping

A primary consideration is to reduce the volume of effluent produced as it is easier to treat a smaller volume of concentrated effluent than large volumes close to the flow rate discharge limit, particularly where the simpler methods involving hydroxide precipitation are to be used. Plating shops are not often over-endowed with floor space, another reason to minimise the volume of effluent treatment facilities.

Tanks must be sound and any drain valves secure and corrosion free. The tanks should be laid out in a logical flow sequence to avoid 'to and fro' actions across the shop floor; they should be located over catch pits or trays which can be periodically emptied and the concentrated spillage recycled or treated separately.

Concentrated solutions from barrels used to plate small items, pipework and in-line filters should be captured during maintenance and not discharged directly but bled very slowly through the treatment facilities. Components should be jigged so that liquid entrapment is minimised and the jigs drained above the tank before transfer.

In manual plant, training operators to reduce carry-over is an essential part of economic operation. Automated equipment should be programmed for the maximum draining times compatible with maintaining work flow rates. A drag-out tank is essential on each plating line. This is a static rinse between the process tank and the first running rinse or swill tank and will reduce metal levels in the final outflow by at least 10 times. The contents can be used to top up the process tank where possible, but as it is essential to keep the drag-out concentration as low as possible to avoid contaminating the subsequent running rinses, batch treatment of the drag-out is more normal and may be required daily.

Returning drag-out contents to process will also recycle contaminants and may be unacceptable where high-quality work to a BS or individual customer standard is in progress. Drag-outs containing chromium are often followed by a neutralising dip which not only reduces hexavalent chromium to the trivalent state but also removes any free acid on the surface of the plated articles.

Further water economies can be made using a counter flow rinse system. Fresh water is fed into the base of the last rinse tank (there are normally three) and cascaded into the second and first swill rinses by taking from the liquid surface and feeding to the base of the next tank. Thus, water flows in the opposite direction to the flow of plated articles. The rinses are usually air-

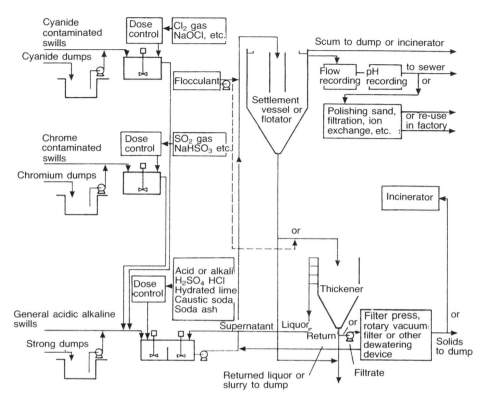

Figure 3.4 Continuous metal treatment plant layout.

sparged and flow controllers fitted to the incoming water, which is often softened first. Compared with a single co-current rinse, such a system can improve rinse efficiency by 20 times and reduce water consumption by five times.

Small dischargers with flows of less than 20 m^3/day may find a batch treatment process satisfactory. Two tanks filled alternately and further reagent storage will be needed with skilled manual operation of valves and pumps. The treated effluent can then be left to settle overnight before discharge. Most factories will have continuous effluent treatment and the possible permutations are shown in Figure 3.4. Note the use of initial dump tanks for each effluent stream. Chemical dosing is carried out by either metering pumps or proportional valves.

3.6 Metal ion biosorption by microfungi

The ability of chitosan, a polysaccharide based on glucosamine, to bind metal ions is well documented. Biosorption is most efficient for those metals that form complexes with ammonia and a stable chelated complex is generated.

Many microfungi contain chitin and chitosan as a structural component of the cell wall and the fibrous nature of the hyphae raises the possibility of creating a mat-like filter for the removal of toxic metals as a polishing stage for industrial effluents. Immobilisation in a polymeric matrix is also under investigation.

Present research has identified *Mucor mucedo* and *Rhizomucor miehie* as able to remove a wide range of metal ions including silver, gold, cadmium, zinc, nickel, copper and chromium. Pre-treatment of the fungal hyphae with sodium hydroxide under controlled conditions followed by exposure to 20 mg/l metal solutions has shown a concentration reduction of 70% in 1 minute and almost complete adsorption within 10 minutes; pH is critical; a pH below 4 causes efficiency to deteriorate markedly. A wash with dilute sulphuric acid is used to strip the metal ions from the hyphae which are reused after pH readjustment.

It is quite likely that future metal treatment systems with very low discharge limits will incorporate some form of biological final clean-up stage.

3.7 Conclusions

Effective reduction of the metal concentration in metal finishing wastes is a simple process chemically but relies on attention to details. These include proper pH control and reliable chemical dosing, adequate settling volume and regular sludge withdrawal. From the author's experiences with plating shops, most of the Consent failures have been caused when the dosing chemicals ran out or desludging had become erratic. More advanced techniques are now being demanded through the tightening of metal Consent limits and this trend will continue. In some cases, zero metal discharge and complete recycle must be expected in the future in many countries.

3.8 Company

BEWT (Water Engineers) Ltd, Tything Road, Arden Forest Industrial Estate, Alcester, Warwickshire B49 6ES.

Chapter 4
Biological Treatment Methods

4.1 Introduction

In order to remove the majority of the organic pollution load in wastewaters, some form of oxidation process involving aeration of the liquid is required. This is normally referred to as secondary treatment. There are two principal methods:

(1) The wastewater trickles by gravity through a filter medium of high surface area relative to volume and which supports a living biomass on its surface, while air passes upwards by natural convection. This biological filter or percolating filter represents a food chain for the living biomass within it, and its successful operation for water purification relies on its being adequately fed and kept wet. The medium can be a graded granular aggregate, and more recently, random plastic shapes have been used widely.

The latter material is often engineered with a very high voidage of 80%−90% which is particularly useful in the main variant of this method, the high-rate filter. These are loaded hydraulically and biologically at 3−15 times conventional rates, and are often used to 'rough' high strength wastes prior to conventional treatment or discharge to the foul sewer as a trade effluent, thereby reducing effluent charges.

Biofilters are always fed with effluent or sewage after primary settlement, to avoid clogging the media with too high a solids load and detritus not trapped by preliminary treatment. Often, a surprising amount of the latter gets through, and rodding out the filter distributor arms is a regular, if not onerous, maintenance task.

(2) In the activated sludge system, either submerged diffusers fed by air blowers, or surface aerators, supply the necessary air and oxygen. About 20% oxygen saturation is maintained to support the metabolism of a suspended biomass termed *activated sludge*, utilising the carbon and nitrogen in the wastewater as a food source. For particularly strong wastes from the food and brewing industries, pure oxygen may be injected or the process contained in an oxygen-rich atmosphere.

A recent variation is an extended aeration and retention system, loosely termed an oxidation ditch, which has some advantages if an effluent low in total nitrogen is required, as treatment includes passage through several anoxic zones where denitrification occurs.

It should be remembered that two significant factors influence the operation of secondary treatment systems, over which little or no control can be exerted. Firstly the climate, which will tend to drive the bio-oxidation reactions at variable rates, and secondly the composition of the incoming sewage or factory effluent. Effluent treatment is one of the few industrial processes where a consistent product is demanded with little control over the raw materials! Good process design with adequate capacity and flexibility are key elements, therefore, in providing a reasonably stable operating environment.

Both biofilters and activated sludge including extended aeration plant will cause a radical, beneficial change to take place to the liquid both in terms of polluting strength and appearance. The two main biochemical processes are carbonaceous oxidation followed by nitrification and the initial reactions are:

$$COHNS + O_2 + bacteria + nutrients \rightarrow$$
(organic matter in wastewater, carbon, oxygen, hydrogen, nitrogen, sulphur)
$$CO_2 + NH_3 + C_5H_7O_2N$$
(new bacterial cells)
$$C_5H_7O_2N + bacteria \rightarrow 5CO_2 + 2H_2O + NH_3 + energy$$
Then: $$2NH_3 + 4O_2 \rightarrow 3H_2O + NO_2 + NO_3$$

Secondary treatment is the only process likely to produce, from an organic wastewater, an effluent of comparable quality to the receiving river or stream water, whilst normally effecting 50%–90% removal of the key parameters – BOD, COD, SS, TOC and NH_3N.

Many plants achieve over 90% BOD removals providing the nutrient balance of nitrogen and phosphorus suits the biomass, and the hydraulic loading rate is correct. Conventional sewage treatment normally achieves 80% or more, while the lower end of the range is represented by industrial effluents containing compounds resistant to biodegradation or where the nutrient balance is unstable and not easy to control. High rate systems are also in this category, but nevertheless are very useful as a preliminary to further treatment, as mentioned above.

One of the principal reactions during sewage treatment is the oxidation of ammonia to nitrate. Analysis for both parameters is quick and relatively simple and 'before and after' samples are often used to provide a primary assessment of how well the process is working. Typically, 20–30 mg/l NH_3N in should be converted to 2–3 mg/l out, with the balance as NO_3N, from an average strength settled sewage. Nitrite levels of 0.05–0.1 mg/l are also characteristic of most filter effluents. Higher levels are often found in winter or following a toxic discharge, as the nitrifying bacteria are more fragile than those effecting carbonaceous oxidation.

Activated sludge usually oxidises ammonia to nitrate slightly more efficiently, being less temperature sensitive. Working out a nitrogen balance is more difficult as the presence of anoxic zones, a deliberate feature of many plant including the extended aeration types, results in the loss of nitrogen gas to atmosphere, and an effluent low in both NH_3N and NO_3N.

Both biofilters and activated sludge are capable of very high ammonia oxidation rates. Farm waste is often over 150 mg/l NH_3N, and there are many examples of sewage works treating this level in admixture with settled sewage at conservative hydraulic loadings, and producing a final effluent of less than 5 mg/l NH_3N.

4.2 Initial considerations for new installations

Assuming some form of secondary treatment is now required to an effluent on a site where previously no treatment or basic sedimentation only was carried out, the following key questions need addressing.

(1) What are the present and future likely effluent standards? Will the choice consistently meet these?
(2) How much space is available, and is it close to housing or other factories? Will there be planning constraints and what is the 'nuisance' level of the plant likely to be?
(3) What budget is available, not only to build the plant but also to run it for many years? What is the energy consumption of the proposed system?
(4) How easily can the plant be adapted to cope with increased flows and higher standards, and can the system be easily tuned and monitored?
(5) Will the plant be mechanically strong and reliable, and able to survive some neglect and overloading?
(6) What level of expertise to run and maintain the plant now and in the future is anticipated or planned for?

Often, local factors will dictate an option or severely limit available choice.

4.2.1 Future effluent standards

Taking these considerations in order, predicting future effluent standards can be a hazardous game, but the short- to medium-term outlook in Europe and the USA is that standards will get tighter, and so some spare capacity and the ability to upgrade must always feature in a design specification.

As the need for high quality effluent production usually accompanies the possibility of refining manufacturing costs through an audit of water and raw material use, product recovery, recycling and energy savings, going for a higher standard than presently demanded by the regulating authority can pay dividends in many manufacturing environments. The key element is to retain some realism; removing the last 5% of solids or BOD can often double running costs and not significantly reduce load to the receiving watercourse, or make but a marginal reduction on the trade effluent bill.

The usual limits placed by the regulating authorities on the effluent from secondary treatment which is followed by a settlement stage are for suspended solids (SS) in the range 10–50 mg/l, BOD in the same range, with often an ammonia standard of 2–5 mg/l. These standards would apply to discharges straight to a river or stream. Both biofilters and activated sludge or their variations are able to produce effluent within these ranges. Discharges to the foul sewer as trade effluents would usually have more lax standards.

In the UK and most of Europe, a 300–400 mg/l limit would be placed on SS, but limits for BOD, COD and NH_3N are less common and normally imposed only where the local sewage works is operating close to capacity. For the same reason, and because of sewer and pumping capacity, maximum discharge rate and 24 hour maximum volumes will also be specified.

Likewise, effluent treatment plant discharging to rivers and streams always have maximum rate and 24 hour total flow limits imposed in the discharge Consent to prevent inadequate dilution generating temporary high solids loads or excessive oxygen demand. Other limits for specific organics peculiar to the process (phenols, oil, solvents, formaldehyde, for example) may be imposed, and will be strict or even nil where the discharge is to a watercourse.

Toxic metal limits will tend to be in the range 1–10 mg/l, but are rarely a problem, as most oxidation systems will not perform reliably above these levels. If the factory effluent contains metals and requires secondary treatment too, metal removal to levels less than 1 mg/l by one of the methods in Chapter 3 is an essential preliminary.

Limits for anions such as phosphate are increasingly imposed at less than 10 mg/l, for which chemical precipitation with aluminium sulphate is often chosen at sewage treatment sites discharging to a water of eutrophic status. Secondary treatment has no significant effect on phosphate levels unless an anaerobic stage is introduced. Some interesting work has been carried out in relation to this (*see* Thomas & Slaughter, 1992).

Meeting these ranges of limits consistently relies on good operational control and maintenance. Above all, adequate settlement and aeration capacity with flow balancing offer the greatest security, coupled with monitoring and alarm on remote or unmanned sites. There is then the chance to rectify mechanical failure or divert a toxic load before it manifests itself in the final effluent.

4.2.2 Available space

Space constraints and the site location often dictate activated sludge or a packaged secondary treatment system. Percolating filters can be a marginal source of odour and the non-biting but very prolific small Psycoda fly is often the cause of vociferous complaints in summer and has a high visual impact. The filters also demand more land area than the alternatives and their potential 'nuisance' value is, therefore, higher under normal operating conditions.

High rate biofilter towers of plastic media used to 'rough' strong industrial organic wastes can be a particular odour source and are not recommended on sites near to housing or offices. Being of high voidage (typically over 80%), they are more temperature sensitive and prone to freezing than the conventional filter. The rotating biological contactor (RBC) is an enclosed version of a biofilter and can be successfully installed in confined spaces with minimal visual and odour impact.

Activated sludge or any mechanically aerated system can be partially or completely screened, buried or otherwise contained, and their operating vices in the 'nuisance' category are usually a result of poor maintenance, and aeration failure when odour problems can rapidly become acute. Their location underneath buildings thus needs special consideration of mechanical reliability, security of power supplies and ventilation to remove gas mixtures, which may contain methane under failure conditions and therefore be potentially explosive.

4.2.3 Budget

Setting a realistic budget for the capital works to construct secondary treatment facilities will need to be based on obtaining some form of guaranteed performance from the plant, a package commonly available from design and build consultancies and construction firms. This implies specifying a quality construction job of high standard where attention is paid to subsequent long-term operating costs and convenience.

The most successful installations often arise from firms offering the complete design, build and commission package. Communication and specification are often better controlled 'within-house' and treatment operating experience is deeper and wider than through a number of separate subcontractors highly specialised in only one or two processes.

The long-term energy requirements of the plant are of primary importance. Electricity is comparatively cheap and very reliable in many countries, but it may not remain so attractive in the long term. Many package plants are energy-hungry, while some designs involve excessive pumping because they were poorly sited or not graded to allow some gravity flows, highly stressed biologically producing a lot of sludge, or require very high recycle ratios to keep media wet. Better balancing facilities would have been initially more expensive but would have entailed much less long-term continuous pumping.

Conversely, the percolating filter installation on a sloping site is particularly attractive. Many rural sewage works use no power at all. All flow is by gravity, and sludges collect in deep wells for occasional tankering. Some sites have no mains electricity connections, any monitoring equipment being solar powered.

4.2.4 Adaptability and the ability to upgrade

Coping with increased flows, standards and necessary tuning in the future can be achieved by building in flexibility to the first design, not only as valves and connections, but also space, for as the factory and the manufactured output grow then so will the effluent volume and strength in many cases. Sufficient room must be allowed on site to add-on extra treatment or settlement tanks. Many package designs come ready plumbed to bolt extra treatment capacity on easily and cheaply.

Initial design flexibility will also allow rearrangements to be more easily made to alter recycle ratios, desludging periodicity, settlement tank retention or dissolved oxygen levels without major reconstruction and cost. Most plant can be tuned with benefit after a period of operation and as a result of operating experience or changes to the effluent composition.

Monitoring can be continuous by electrodes or sensors for out-of-range alarm conditions, or through formal analysis in the on-site laboratory. The benefits and resourcing of this is explored in Chapter 7.

4.2.5 Reliability

Mechanical strength and reliability are essential in wastewater treatment equipment in view of the arduous and continous operating needs. Fortunately, typical components for the water industry are well made by established manufacturers and suppliers and enjoy a long service life.

When specifying equipment, capital cost should not therefore be a prime consideration. More important are long-term reliability, life expectancy, the cost and availability of spares and the manufacturer's track record.

4.2.6 Expertise

The level of expertise needed to run and maintain an industrial effluent or sewage treatment plant over a long period of time is frequently underestimated. There have been some expensive mistakes with turn-key projects around the world where the level of training and understanding by the local operators has quickly proved inadequate to cope with the plant, the latter often being over-sophisticated and lacking in spares. Adequate training is therefore fundamentally important and should be started while the plant is being constructed.

Personnel need to be of adequate calibre to understand at least the basis of operation, how to assess correct functioning, the importance of reporting and attending to plant equipment failures, and the legal consequences of discharges outside limits. Plant attendants are often very enthusiastic characters, but some of the new systems require a high level of understanding and technical expertise to diagnose and rectify problems. Formal academic qualifications or considerable experience are often essential and the effluent treatment plant manager or supervisor should be regarded as having equal status to the site production manager. These areas can be addressed by a formal programme of training with refreshers as necessary and the provision of adequate supervision. The following are essential good practice:

- a quality management system with flow diagrams;
- the labelling of all components and valves to aid identification;
- instructions on making up chemical solutions, and
- a 'What to do if' checklist.

If something goes seriously wrong, a system is then in place to demonstrate organisation and competence to the would-be prosecutor.

4.3 Some operating practicalities of biological oxidation

The micro-organisms responsible for biological oxidation operate at an optimum pH range of 7–8.5 and a nutrient balance in the ratio of BOD:N:P of 100:5:1. Some industrial effluents will therefore require nutrient addition and pH adjustment consistently and accurately to achieve these ratios before entry to biological treatment, and this is vitally important if successful operation is to be achieved continuously.

Other limiting elements are K, S, Ca, Mg, Na and Fe. Micronutrients include Zn, Mn, Mo, Se, Co, Cu, Ni, V and W.

Hydraulic and biological loading rates require to be maintained within certain limits. There are no hard and fast rules; from the many papers written on secondary treatment comes evidence of plant producing the same effluent quality from loadings that are 2- to 3-fold different. Further guidance is given in the sections below that describe the differing processes more fully.

As any living biological system strives to maintain a steady state and is relatively fragile, wide excursions from the ratios above, sudden shock loads and toxic materials will rapidly impair performance which will be slow to recover as the biomass population takes a number of generations or 'ages' to recover to original numbers.

Additional operational problems, apart from a poor effluent caused by low oxidation, include odour from a dead and decomposing biomass, and the difficulty of disposing of a highly putrescent and possibly toxic sludge. Considerable time and expense can be involved. A biofilter system seriously poisoned or left to dry out will often take 3–6 months to recover fully at ambient temperatures between 10°C and 30°C, and longer in cold climates. Recovery is best speeded by effluent recirculation. Activated sludge plant is best reseeded with sludge from another which is operating properly, and should then stabilise within one month. Complete replacement of all the activated sludge may be necessary. The costs of these measures in time, manpower and tankering can be considerable. Meanwhile, the effluent will continue to flow in for treatment!

A fundamental requirement for biological treatment of industrial effluent is therefore to assess biological treatability over the range of effluent types that the factory is likely to produce and build sufficient monitoring, security and balancing into the system to avoid shock or toxic loads ever reaching the biotreatment system.

It is equally important to maintain flow to keep the media wet and the system 'fed' with some form of nutrient input. Recirculation is thus virtually essential for filters to cover periods of low or no flow, typically at night on most sewage works. Similarly, activated sludge is always returned to the system as part of the process to maintain the biomass.

In most circumstances, the rate of biological reactions doubles with a 10°C rise in temperature. Biological oxidation works better in summer than winter therefore, and some allowance for this is often made in effluent discharge standards where higher levels of BOD and ammonia are permitted for the colder months.

Bio-filters, and particularly the high-rate variety with high voidages which act like cooling towers, exhibit this decline of oxidation performance (of about 10%−20%) more than activated sludge systems, where the exposed liquid surface is low in relation to tank volume. Inflowing wastewaters including sewage are often above ambient temperature, however.

With few exceptions, primary sedimentation to remove some of the solids load precedes secondary treatment, and without exception, some form of pre-screening to remove gross material is essential.

After the oxidation stage, another stage of settlement is essential to separate the oxidised and usually highly active sludge from the effluent. The humus sludge from filter effluent is normally mixed with the raw sludge from primary sedimentation to facilitate dewatering. It rarely dewaters by itself to more than 4% DM, and has an earthy rather than an offensive odour. Filters seasonally 'slough off' solids in spring and autumn as the ambient temperatures change, imposing a temporary increase in solids loadings to tanks and often a temporary rise in effluent BOD and SS values.

Activated sludges from mechanically aerated or diffused air systems are recycled back to the process and some wasting is necessary in most situations. They are rarely above 2% as removed from settlement tank(s) and are difficult to dewater above 3% DM, even with chemical dosing. Both activated and extended aeration systems produce plenty of sludge − often 50% more than the solids load entering the aeration plant. Adequate storage and de-watering capacity must therefore be allowed; it is a common design defect to underestimate sludge production from these systems. Chapter 5 reviews sludge treatment techniques.

Both humus and activated sludge will denitrify and float readily in warm weather, causing sludge loss over the weirs. Secondary settlement tanks are therefore always desludged regularly (every 1−2 h) or continuously, except on the smallest filter works where once a day should suffice. Total nitrogen content of humus and activated sludge is about 3% DM, while extended aeration systems may produce sludge up to 8% nitrogen on DM, much of it trapped in the liquid phase.

To avoid floating sludge and tank inversions, it is not therefore normal practice to return secondary sludges or surplus activated sludge high in nitrogen for co-settlement with primary raw sludges, where the reducing conditions will cause rapid denitrification and such arrangment must be operated very cautiously. The author has seen this work on some sites, not on others. The high soluble nitrogen content of the sludge may be the determining factor, making the sludges from extended aeration particularly unsuitable for co-settlement with primary sludges.

4.4 Biological oxidation microbiology

With few exceptions, the micro-organisms associated with wastewater treatment do not persist through the whole process or even the length of the sewerage system, being relatively fragile and therefore colonising only those areas attractive to their individual requirements.

Both biofilter and activated sludge systems represent a food chain, and the range and type of organism will be different in each. Frequently present in activated sludge are *Zoogloea, Achromobacter, Flavobacterium, Nocardia, Mycobacterium, Bdellovibrio, Nitrosomonas, Nitrobacter, Beggiatoa, Thiothrix* and *Geotrichum*. Biofilters also host nematodes, insect larvae, microfungi, algae, small worms and the freshwater leech.

Microbiological examination can identify changes to relative concentrations of the organisms at times of stress and is useful in identifying the cause of changes to effluent quality. For example, a cause of bulking and difficult settling activated sludge is a preponderance of filamentous organisms including *Microthrix* spp., induced by, amongst other factors, high concentrations of carbohydrate in the feed water. Nevertheless, most monitoring of bio-treatment is carried out by chemical testing of the wastewater influent and effluent quality.

The activity of the organisms is assessed by measuring the effect of the oxidation reactions within the system. Reductions in BOD and the oxidation of ammonia to nitrate are particularly important parameters. Two basic oxidation reactions occur:

(1) Carbonaceous oxidation of a very wide range of compounds containing C, H and O, including toxic ones such as cyanides, phenols and oils in low concentrations. The end products are theoretically carbon dioxide and water, but complete oxidation of all components never occurs, and the effluent always retains some partially oxidised material measured as BOD or COD.

(2) Nitrification, where the organic nitrogen present in sewage or wastewater as proteins, amines and urea are hydrolysed to ammonia, which is oxidised

to nitrate. Two organisms, *Nitrosomonas* and *Nitrobacter*, are involved and the reactions proceed in two stages thus:

Nitrosomonas:
$$55NH_4^+ + 76O_2 + 109HCO_3^- \rightarrow C_5H_7O_2N$$
$$\text{(new bacterial cells)}$$
$$+ 54NO_2^- + 57H_2O + 104H_2CO_3$$

Nitrobacter:
$$400NO_2^- + NH_4^+ + 4H_2CO_3 + HCO_3^- + 195O_2 \rightarrow$$
$$C_5H_7O_2N + 3H_2O + 400NO_3^-$$
$$\text{(new bacterial cells)}$$

In order to oxidise 1 mg of NH_3N to NO_3N, 4.3 mg of oxygen is required. The reaction consumes a large amount of alkalinity; 8.64 mg as HCO_3^- is required to oxidise 1 mg of NH_3N.

While carbonaceous oxidation occurs first and the autotrophic heterotrophs are relatively robust, so that some oxidation will usually occur unless the system is completely poisoned or dead, nitrifying bacteria are more demanding of the optimum conditions of a thin, mostly aerobic biomass. In particular, the pH range of the wastewater must lie between 7.3 and 8.6.

Biosystems which are heavily loaded organically and support extensive film growth, or in the case of activated sludge or extended oxidation are operated at high solids levels, will undergo effective carbonaceous oxidation but will not nitrify. The ammonia content of the effluent will then be similar to the plant influent.

This state in sewage treatment indicates overloaded conditions or inadequate aeration, but may be acceptable in industrial scenarios where nitrification is actively inhibited by some materials in the flow to treatment. The best filter medium is therefore well graded to secure good ventilation and has a rough, craggy surface with plenty of interstices to encourage the floc-forming *Zoogloea* types of rod-shaped bacteria to grow. A 'zoogloeal slime' over the media surface is a healthy sign, but since the filter is a food chain, the presence of insect larvae, worms and mosses should also be expected. Only the top 1 m of media has any appreciable slime, the lower depths of the filter being populated by nitrifying bacteria.

A number of fungi and algae also develop, the latter consuming carbon dioxide and oxygen and utilising light, and are thus restricted to the surface. Birds like starlings and gulls will frequently pick over the filter media for worms and insect larvae. Fungi perform a similar function to the bacteria, but excessive numbers in activated sludge hinder solids settlement by reducing density and causing 'bulking'. They often indicate an incorrect C:N ratio and high carbohydrate waste liquors are prone to this.

The mechanism of BOD removal in a filter or an activated sludge biomass is complex. Effluents will contain pollutants (the food source) in solution or as suspended solids, and the major rate-limiting step is biophysical adsorption and absorption by the biological film.

Subsequent bacterial metabolic stages are numerous and ordered by reaction rate which is temperature sensitive. Solids are broken down by extra-cellular enzymes and dissolved material adsorbed into the cells of the biomass. Growth and endogenous respiration is followed by excretion and grazing by organisms like protozoa, and in filters, insect larvae, worms and birds, a food chain extending to higher organisms.

Rate-limiting steps in these processes are the oxygen transfer rate from the aqueous phase through the cell wall, and the actual oxidation rate within the cell. Although the equilibrium saturation value for oxygen in water is between 8 and 11 mg/l over the temperature range of most liquid effluents, the rate of bio-oxidation is not adversely affected until less than 1 mg/l is available, and activated sludge and extended aeration plants are often run with only 1–1.5 mg/l dissolved oxygen available.

Endogenous respiration, where bacteria will oxidise their own cell material, is induced by a reduction of BOD load (the food source) to the system and tends to lead to temporarily poorer effluent, as the biomass is reduced in both numbers and metabolic readiness when load is re-applied. Thus, recirculation or flow balancing to maintain reasonably stable conditions by returning viable organisms and nutrients is highly desirable.

Recirculation in a filter plant also serves the function of returning fine particles to the filter to assist in floc formation by agglomeration. The polysaccharide sheath secreted by the zoogloea and Pseudomonas-derived bacteria is the site of this activity, involving a reduction of electrostatic surface charge, and the removal of much colloidal solid material as flocs. As floc production relies on particle collision, the process is second order kinetics, and proceeds slowly in the dilute suspensions of a liquid effluent. Recirculation returns fines to the system and increases the chances of collisions and floc formation. This process is essential for good settlement, well defined sludge/water boundaries in the final tanks and a clear effluent.

Bacterial removal rates for both biofilters and activated sludge plant are typically 95%–99%, these being design values frequently used at present to assess bacterial levels in marine discharges and likely compliance with the urban waste water Directive (91/271/EEC) after dilution. High-rate biofilters are often quoted at 50%, but this will be different at each plant, and very dependent on the effluent composition.

4.5 Secondary treatment plant types

4.5.1 Conventional biofilters

Also termed percolating filters, the standard clinker bed filter is a common sight worldwide, having formed the foundations of sewage treatment since its initial development in the 1890s by Dibden and Clowes. They were first used on a large scale in 1893 in the UK at Salford STW.

The usual filtration media is blast furnace slag, although gravel, shale, coke, coal and rigid plastic shapes have also been used with success. About 2 m deep, and with the media carefully graded in the size range 25−150 mm diameter with the largest material at the bottom, the whole bed is constructed on a system of underdrains. This allows a free flow of air to pass upwards through the filter by natural convection as effluent drains downwards by gravity. Retention time within the filter media is typically 20−35 minutes to achieve 80%−90% reduction. High-rate filters have rarely more than 10 minutes retention, hence the reduced removal performance.

Liquid is conveyed to the media by hollow metal distributor arms with regularly spaced holes and supported by wires. On circular filters, the jetting action of the liquid leaving the arms drives the distributor round. Fishtail spreaders are sometimes fitted to spread the liquid more efficiently over the filter media. Circular filters, therefore, require no external energy source, providing the central distributor bearing is free and the arms do not drag on the media at any point during rotation.

On exposed sites, wind vanes have been used to assist rotation in adverse weather; at sites where much of the flow is pumped and highly erratic, motor drives are fitted to the distributor arms principally to even out rotation speed variations.

Circular filters vary in diameter between 4 and 20 m. Figure 4.1 shows a

Figure 4.1 Close-up of the distributor arm of a circular percolating filter.

Figure 4.2 Circular biofilter set.

close-up of the distributor arm of a circular percolating filter, and Fig. 4.2 a typical sewage works filter set. The rectangular filter is more economical in ground area, but requires a motor driven wire winch system to draw the distributor arms up and down the filter bed. This layout needs careful alignment and maintenance, and is thus confined to larger, manned sewage works.

Figure 4.3 shows the compact nature of a rectangular bed and the winch wires.

As already mentioned, biofilters are only fed with settled sewage or industrial effluent after initial settlement, as a high solids load will quickly block the distributor arms and clog the top media surface. Effective preliminary treatment to remove grit and rag is also essential. The filter is either fed continuously from a distribution chamber at a slightly higher level than the distributor arms, or via dosing syphons on a cyclic basis at sites where the flow is too low to maintain filter arm rotation, or much of the influent is pumped periodically.

Figure 4.3 Rectangular biofilter set.

Maintenance of biofilters is confined to cleaning out the distribution holes in the arms and rodding out the whole arm (Figure 4.4) – a daily task if solids levels are over about 250 mg/l – and occasional retensioning of the support wires. Many filters can be maintained on a weekly check basis, however, and the simplicity of design has encouraged their use in remote, unmanned areas.

Correct filter rotation can be sensed remotely by telemetry, using a proximity detector mounted on the peripheral edge of the filter bed, and triggered by a metal bar on one of the arms. Rotation detection on small sites where the filters are periodically dosed and are therefore stationary for much of the time requires a more 'intelligent' approach. Warren Jones Engineering supplies a monitor relating anticipated rotation to flow, which can be programmed differently to account for the hydraulics at any site and will monitor up to four filters.

Biofilter performance is dependent on a number of factors. The environmental parameters of ambient temperature, over which little control is possible, and organic loading applied, are established principal factors affecting biological removal efficiency.

Winter/summer performance differences in biofilters can be considerable. A 25% reduction in nitrification and 10% in BOD removal should be allowed for a temperature drop of 15°C. Partial compensation for lower temperatures can be achieved by reducing the hydraulic and organic loading rates and so increasing retention time – frequently not practical where the

Figure 4.4 Rodding out a filter arm.

rainfall and flows are higher in winter. Assessment of filter performance can thus only be realistic if carried out over the complete cycle of seasons. There may be local, seasonal or production factors which have a pronounced short-term effect too.

The two media design variables that most influence biological treatment efficiency are the specific surface area available for biological growth, and the void space within the filter, allowing ventilation and drainage. For mineral media, the former requirement is most effectively served by blast furnace clinker material which remains much sought after in this application. 28 mm lumps of slag average $220\,m^2/m^3$.

Random packed plastic has many advantages providing it can be adequately wetted. This often requires a high hydraulic loading rate, hence the use of plastic in high-rate filters as discussed later. Plastic media offers a high voidage (80%–90%) and low retention time, not always suitable for sewage treatment where the maximum possible oxidation of a relatively low strength organic waste water is sought. Hemmings and Wheatley (1979) give an account of an investigation of plastic media use in sewage treatment while Hill et al. (1992) provide details of an unusual application of high rate filtration specifically for nitrifying the effluent from an aeration plant.

Much of the poor or ailing performance of older installations arises from collapse of the underdrains and degeneration of the media by frost damage and erosion. Filters with exposed sides and built at ground level are more prone to frost damage, and can freeze solid in very severe winter weather, whilst those buried in the ground are better thermally insulated but often suffer from restricted air flow. The author has never found any significant performance differences at normal operating temperatures that can be specifically attributed to construction between these differing installations.

Nevertheless, the life expectancy of filters is often over 50 years, and even then, refurbishment of the distributors and regrading of the media will often restore performance to original.

Conventional biofilter loading rates

Biofilter hydraulic loading rates are usually quoted at $0.60\,m^3/m^2/day$ to achieve 90% nitrification, with $0.8\,m^3/m^2/day$ as an upper limit for 80% removal of BOD and ammonia. Whilst acting as a useful guide, hard and fast rules are not appropriate as biofilters can be highly individual in performance, even within the same site and fed with the same feed wastewater. There are many examples of well graded, walled filters achieving 90%+ removals at loadings approaching $1\,m^3/m^2/day$ where filter voidage and drainage rate and available media surface area are all high. Conversely, hydraulic underloading leading to drying out and a biomass starved to the point of endogenous respiration will not produce good effluent quality either.

This can be a hazard on small works, or industrial sites where flow is erratic and sometimes biologically very strong for short periods. Flow balancing, continuous recirculation and double filtration (where there is more than one filter bed) are all solutions. The main message is to design conservatively for hydraulic loading, but not too conservatively! For sites where volume increases are in the future, build more filter capacity than needed, but avoid the temptation to use it until flow rates justify.

A biological loading rate of $0.1\,kg\,BOD/m^3/day$ is frequently quoted to achieve full nitrification; 80% BOD removals may be expected at $0.2\,kg\,BOD/m^3/day$ and represents the boundary between low and high rate filtration. A poor ammonia removal of less than 50% may be expected at this rate. The British Royal Commission Standard, now of some vintage but a well tried design parameter, is $0.12\,kg\,BOD/m^3/day$ to effect the highest BOD removal and nitrification.

Effluents high in ammonia but low in BOD should be applied at the rate of $0.05-0.08\,kg\,NH_3N/m^3/day$, and conventional hydraulic rates. This can only be a rough guide, as the BOD/NH_3N ratio and the C:N:P ratio will also affect removal rates; $0.08\,kg\,NH_3N/m^3/day$ is likely to achieve about 60% removal.

Percolating filters will adapt to the treatment of toxic materials over a period of several months providing the concentration is increased very slowly. There are thus many documented examples of the effective oxidation of phenols, cyanides, pharmaceuticals and non-chlorinated solvents and toleration of metals without significant inhibition at the 10+ mg/l level. Important factors are the maintenance of adequate flow and a reasonable nutrient balance. Once acclimatised, the biofilter system will usually tend to tolerate shock loads of toxic materials and high BODs rather better than the activated sludge system.

4.5.2 High-rate biofilters

High-rate filtration has been defined historically in the UK by the Department of the Environment as that where the hydraulic rate exceeds $3\,m^3/m^2/day$. Development took place in the early 1960s as rigid plastics became available, and it was realised that this form of filtration had considerable potential to 'rough' high-strength industrial effluents prior to sewer discharge or conventional low-rate treatment. The food and chemical production industries have found this technique particularly attractive, where effluents often have high soluble BODs and CODs in relation to solids. A considerable reduction in trade effluent charges by reducing the strength factor has also proved an attraction.

As the media are inherently light, high-rate towers are typically 3–4 m high, often rectangular, almost always entirely above ground and enclosed in lightweight metal sheeting. Earlier installations may well be clad with corrugated asbestos, the handling hazards of which should be borne in mind should media replacement or modifications be contemplated. Figure 4.5 shows such a tower treating a food waste. As plastics are not easily wetted, and the

Figure 4.5 High-rate biofilter tower.

voidage of high-rate filters is frequently 90%, irrigation paths and rates are crucial to good performance. Retention times are rarely more than 10 minutes. Despite this, 50% removal rates of both chemical and bacterial numbers should be achieved.

The media must be randomly packed during construction to avoid short-circuiting, and spray irrigated at a rate to ensure continuous wetting. Nozzle design is important in achieving a good surface spread of liquid, whilst resisting blockage by high solids or colloidal material that may coalesce with temperature change. Figure 4.6 shows a well-engineered distribution system that has worked successfully for over 20 years with an effluent containing fat particles.

In common with conventional rate filtration, a preceding settlement stage is essential to capture gross solids, even if SS levels are lower than the BOD values.

Important chemical properties of the plastic media include inertness to biological degradation, resistance to attack by traces of organic chemicals (particularly solvents in the effluent), and a resistance to degeneration by ultra-violet light or deformation by temperature changes. Extended life tests have indicated the last two factors to be less significant than expected due to the protection offered by the biological film, and most manufacturers expect a 50-year life for plastic media.

Mechanically, it should be sufficiently structurally rigid to bear its own

Figure 4.6 Distribution system to high-rate filter.

weight and that of overlying media together with the attached biological film. This is particularly important, as degeneration in performance is often due to compression and collapse of the bottom layers of media, remembering the height of some filters.

Suppliers of media are numerous. BS Floccor Ltd and Mass Transfer International are two that offer a number of media shapes to treat different types of effluent over a range of loading rates.

High-rate biofilter loading rates

Hydraulic loading rates of $5-10\,m^3/m^2/day$ are common and $25\,m^3/m^2/day$ possible. This often requires pumping the effluent onto the filter under pressure or maintaining a high recirculation/flow ratio (5–10 times) and the electricity costs of this should not be underestimated. Biological loading rates often exceed $1\,kg/m^3/day$. Again, individual circumstances will dictate performance. Hill *et al.* (1992) quotes $0.08\,kg\,NH_3N/m^3/day$ as achieving 60% removal, although the high-rate filter is normally selected to perform carbonaceous oxidation only rather than nitrification, and achieving the latter is often an erratic bonus at the designed hydraulic rates.

Several major attractions of the high-rate filter include the low construction costs and rapid installation. Ground preparation need not be extensive. Surface loads will be low as the media is so light, and excavation apart from drainage unnecessary. The system can be up and running in a few weeks.

Toleration to shock loads, loss of feed and drying out are similar to conventional filters, and many high-rate filters operate on a surprising cocktail of organics. Recovery is normally quicker, because of the high irrigation rate and the fact that normal performance comprises only carbonaceous oxidation which re-establishes much quicker due to the robustness of the organisms, than the more fragile nitrifying bacteria *Nitrosomonas* and *Nitrobacter*.

Several less positive factors also arise from low residence times, and high irrigation rates and voidage.

(1) Spray drift may prove a nuisance at high rates of application, and in strong winds. In view of the nature of many high BOD wastes, this can give rise to an unpleasant working environment near the biotower.
(2) The high voidage tends to make the filter act like a cooling tower and treatment efficiency rather sensitive to ambient temperature. Freezing up in extreme winter weather may not be prevented by a warm effluent. Coincidently, a lot of water vapour will be emitted from the top surface, not unattractive but a surprising source of complaint in built-up areas.
(3) A high-rate tower can be a particular source of odour arising from the high ventilation rate and effluent strength. They provide the ideal breeding ground for the Psychoda fly, non-biting and short-lived, but unpleasant by sheer numbers. Both these factors dictate siting the tower well away from housing, offices and even roads if you want a quiet life!
(4) Sludge production can be higher than conventional filters, due to partial oxidation, and it may prove more difficult to settle or dewater than a conventional, well-oxidised material. The sludge is also likely to be both a considerable source of odour and regarded as a primary sludge, restricting disposal routes within EU countries.

Whilst maintenance of the actual high-rate filter is negligible and confined to clearing spray nozzles, running costs where high recirculation rates to maintain flow and wetting are required will be higher than conventional biofilters and skilled maintenance staff needed to carry out repairs to ancillary pumps and valves.

4.5.3 Rotating biological contactors (RBC)

First installed in West Germany in 1960, this 'package' system of biotreatment has been used more for carbonaceous oxidation and as a roughing filter than in a nitrifying mode, if the majority of papers is any guide to applications.

A series of discs or drums of plastic media slowly rotate on a horizontal shaft, being alternately exposed to air or dipped in a trough containing the wastewater. Close-packed discs on an 8.2 m long shaft offer $9000 \, m^2$ of surface area while corrugated high-density polypropylene discs offer up to $16\,000 \, m^2$. Typically, 40% of the biofilm that develops on the discs is submerged, although Rosewater Engineering supplies a package plant with plastic segments forming a three-dimensional open tube which is 70% submerged.

Figure 4.7 shows a cross-section of a modular design supplied by Klargester Environmental Engineering Ltd and containing primary and secondary settlement zones. Figure 4.8 is a small unit treating domestic sewage from four houses and taken from the motor drive end. The RBC is fed with settled sewage or industrial effluent, the modular approach offering ready expansion and plants up to 30 000 population equivalent have been built.

Typical hydraulic loadings of $0.08-0.16 \, m^3/m^2$ of disc surface area/day and organic loadings of 0.01 kg soluble BOD/day achieve 60% BOD removal. A second unit is added to give nitrification where required. These loading rates normally produce a 40/30 effluent after settlement. RBCs are particularly

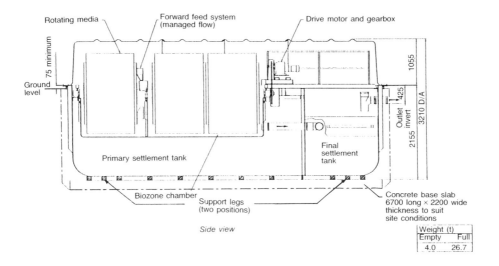

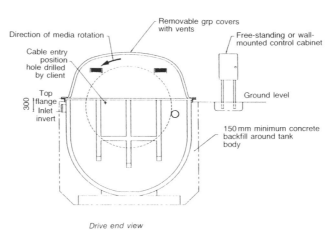

Fig. 4.7 Cross-section through an RBC. (All dimensions are in mm). (Courtesy: Klargester Environment Engineering Ltd.)

Figure 4.8 Small RBC unit.

useful where a low headloss is demanded, and being covered, where fly and odour nuisance precludes open percolating filters.

The 'package' nature of the plant has led to their use for small communities or isolated factories with organic wastewater to treat, assuming there is power available. The plant occupies little room and as half is buried in the ground, is visually unobtrusive.

Operationally, RBCs tend to be less demanding than biofilters with rotating arms, having no dosing syphons or jets to block up and being mostly unaffected by freezing conditions. Routine maintenance is confined to greasing bearings and drive components.

Problems with RBCs are normally centred on organic overloading. Excessive biofilm development takes place, bridging the media and generating anaerobic conditions. High shaft loadings occur as the media discs acquire a thick biofilm causing shaft and bearing collapse. After a long power failure, unequal biofilm build-up can cause the drive mechanism to stall and break the chain drive. It is important therefore to keep the soluble BOD load applied as low as possible by regular and thorough desludging of the settlement zones. A shear pin in the drive chain and loss-of-rotation sensing coupled to an alarm are other refinements.

4.5.4 Activated sludge

This process was first devised in about 1913 by Ardern and Lockett and tried out on various sites in Northern England. It did not become the established process on all the largest sewage treatment sites until the 1950s but has since superceded biofilters at such locations for several reasons.

(1) The process is less land consumptive.
(2) Odour is less and the Psychoda fly absent, important factors when the works is close to housing and factories. Pressure for land in many parts of the world has resulted in many sewage works sites, originally out in the country, becoming surrounded by development.
(3) Although requiring an energy input, most large sites generate electricity from the anaerobic digestion of raw sludge, often with excess to spare and technically able to export this to the electricity grid. Thus, through the operation of 'green tariff' incentives, the energy costs of running the aeration plant can be nil.
(4) The effluent tends to be of better and more consistent quality, as the process is temperature independent.

Treatment takes place in a large tank in which a microbial floc – the activated sludge – is maintained in suspension by mechanical surface aeration or diffused air blown in under pressure from the tank floor.

The microbial floc comprises flocculated unicellular bacteria – both spherical and rod-shaped – competitively co-existing with filamentous types, actinomycetes and fungi. Protozoa, metazoa and nematode worms complete a diverse ecosystem.

Activated sludge is perhaps the only wastewater treatment process worthy of microscopic study when the process is not working well, and some of the common problems arising from an imbalance of these organisms are described below.

Establishing the activated sludge floc from settled sewage when commissioning a new plant is a fairly lengthy process needing high levels or aeration and mixing that may take a month or two, and the water industry usually 'seeds' the process by tankering in surplus activated sludge from a nearby operating site. Often, they are willing to assist industrial plant commissioning for nothing, or the price of the tankering if some distance is involved.

The mixed liquor suspended solids (MLSS) is maintained at between 2200–3000 mg/l in most plant where nitrification is required. To achieve this, and bearing in mind that the influent contributes a constant solids load and that the organisms are constantly reproducing and generating solids, some sludge wasting from the system is always required. A 50% production of sludge within an activated sludge plant is not unknown. Adequate surplus sludge consolidation and handling must be allowed for; many present plant designs display a lack of appreciation of this fact through inadequate sludge storage facilities, and present constant operating difficulties.

Tanks vary considerably in size, typically having a length/width ratio of between 4:1 and 10:1 and averaging 3 m deep. Some plant consist of a series of lanes round which the liquid flows, each lane being 30 to 50 times longer than its width.

As with biofilters, the feed is settled sewage or industrial after some preliminary settlement. After much experimentation over the years, better operating performance has been gained by feeding the influent and returned activated sludge (RAS) from the subsequent settlement stage in at one end of the plant, creating plug flow conditions through the tank, rather than completely mixed. Aeration levels can then be controlled at different points and anoxic zones generated.

Influent and effluent flow continuously, the latter over a weir to one end or side of the tank and which is capable of height adjustment. This feature alters the top liquid surface level in the tank, controlling the degree of immersion of mechanical aerators if fitted, and their air input. Weir height control is therefore linked to preset dissolved oxygen (DO) levels. Liquid retention time is usually 6–8 hours.

Aeration equipment

Aeration by diffused air is achieved by remotely sited, soundproofed air blowers supplying a large number of diffuser domes on the tank floor with air at sufficient pressure to overcome the head of liquid. Some form of filtration to remove oil vapour and reduce water vapour is desirable, and in cold climates pipe lagging to avoid freezing of valves and actuators.

Both fine and coarse bubble domes are available; the former are much more common. The domes are threaded into plastic supply pipework, so facilitating removal for cleaning after a period of time, usually about 5 years. Cleaning may require acid to remove scale, and firing at 600–800°C to burn off organic deposits. Some 5% loss by breakage should be allowed for. Clearly, the tank design will have to include valving facilities to isolate and drain sections or lanes for maintenance of this type of aeration. An opportunity is also presented to clean out the grit and detritus that eluded preliminary treatment and that will have settled out, reducing treatment capacity.

Mechanical surface aeration is achieved by cones rotated by a motor/gearbox combination at about 60 rpm and fixed to a superstructure (Figure 4.9). Typical motor ratings are 12–40 kW. Horizontal rotors fitted with metal 'brushes' can be used, but are more common on extended aeration systems.

Dissolved oxygen (DO) control

DO levels in activated sludge systems are best maintained at between 1.5 and 2 mg/l for reasons connected with the metabolism of the types of organism encouraged and discouraged at this level and the settling characteristics of the activated sludge. Some form of control is essential as most influent will vary in strength and oxygen demand over a 24-hour cycle, and in industrial environments often suddenly and over a wide range.

Figure 4.9 Surface aerators – activated sludge.

Depending on plant size and layout, one or a number of dissolved oxygen electrodes will be required, the maintenance and regular (weekly) careful calibration of which is essential by trained staff. Most electrodes are quite large and robust and will operate for over 6 months before a service is needed, consisting of light abrasive cleaning of the silver anode and replacing the membrane on the Mackereth type. ELE International Ltd supplies the water industry widely.

Calibration at 0% and 100% saturation is achieved by electrode immersion in 5% sodium sulphite solution for two minutes, followed by gently swinging in air. An electrode in good working order should read 95% of true value within 45 seconds.

To control DO levels by diffused air, blowers may be switched on and off, or more elegantly, at least one variable speed blower is included and controlled via a loop and from the signal output from one or several DO electrodes. Dividing the air supply to the diffusers so that some can be turned off is less strongly recommended because blockage often results which the air supply will not always clear on reinstatement, and also non-return valves will be required if the flooding of unpressurised pipework is to be avoided.

A recent development consists of a fine bubble system with diffusers mounted on a flexible rubber membrane to form an aeration mat. This can be suspended in a tank whilst the permanently installed equipment is being refurbished, for emergency aeration, to cover DO deficiency caused by high BOD shock loads, or to uprate a plant on a semi-permanent basis.

More choices are possible for DO control with mechanical surface aerators.

(1) Several may be switched on and off where there are a number of them. This may, however, lead to a loss of mixing and dead pockets.
(2) Rotation speed may be altered: elegantly and relatively efficiently through electronic control; mechanically by electrically actuated variable ratio gearboxes. The latter method is less efficient but often more reliable on sites where mains voltage spikes and fluctuations may cause havoc with electronics.
(3) DO may be controlled mechanically via a rise and fall gearbox so that the degree of aerator immersion is altered. These have proved to be reliable but high maintenance items on many sites.
(4) The effluent weir may be driven up and down by a proportional actuator. This is simple and responds quickly to DO changes by varying top water level in the aeration tank and therefore aerator immersion depth. Turning a number of aerators that are barely skimming the liquid surface is costly electrically, but the system is widely used on extended aeration with horizontal rotor aerators.

A benchmark for aerator efficiency is that the system chosen should dissolve 2 kg oxygen/kWh used. Upon commissioning a new plant, a common way of checking this is to fill the whole aeration plant with water or final effluent, and add sufficient sodium sulphite to reduce the DO level in the water to zero, as measured by DO electrode.

Clearly, several tonnes of the chemical will be required in a plant of any significant size. The aeration system is then started, and the time taken and kWh consumption measured to achieve a set DO level in the tank or lanes. This can be compared with the theoretical amount of oxygen required to oxidise the sodium sulphite to sodium sulphate, and raise the DO from zero to, say, 2 mg/l for the liquid volume in the system.

Operational plant aeration systems often fall short of the target of 2 kg/O_2/kWh and 1.3–1.6 kg/O_2/kWh are more typical values, bearing in mind the variation of immersion depth of mechanical surface aerators will be used to control DO levels in many plant.

Manufacturers of surface aerators typically claim oxygenation capacities of approximately 2.4 kg/m^3/h at 125 mm immersion, rising to 3.7 kg/m^3/h at 200 mm immersion.

Activated sludge design parameters
(1) The organic loading rate or food to biomass (F/M) ratio expressed as kg BOD/kg MLSS/day. This is calculated from the formula:

$$F/M = \frac{So - S}{Ax}$$

where:

So = influent BOD in mg/l.
S = effluent BOD in mg/l.
A = hydraulic retention time in the aeration tank.
x = concentration of volatile SS in the aeration tank in mg/l.

A typical value is 0.3; carbonaceous oxidation with little nitrification will occur at rates of 0.4−0.6. For full nitrification, 0.1−0.25 is recommended. Fill and draw systems and extended aeration may operate at 0.05−0.15, but in these the MLSS levels are higher—typically 4000 mg/l.

(2) The volumetric loading rate, expressed as $kg BOD/m^3/day$. A typical value for conventional activated sludge is 0.5. Many documented plant operate satisfactorily at 1.0.
(3) The retention time of the biomass within the system or sludge age. On most plant, 10−20 days is the norm.
(4) Installed aeration capacity measured as $kg O_2/m^3/day$. Between 1.0 and 3.5 is usual, allowing for some upgrading.

Oxygen availability is an alternative way of expressing aeration capacity, typically $2 kg O_2/kg MLSS/day$ for full nitrification. Oxygen requirement is based on the level of treatment required, expressed as $kg O_2/kg BOD$ removed. Carbonaceous oxidation is achieved by $1.0-1.2 kg O_2/kg BOD$, but nitrification needs 1.6+ on most plant documented in the literature.

Oxygen requirements for an F/M ratio more than 0.3:1 are $30-55 m^3 O_2/kg BOD$ removed for coarse bubble diffused air, $25-35 m^3 O_2/kg BOD$ for fine bubble. At lower F/M ratios, where nitrification must be guaranteed, $75 kg m^3 O_2/kg BOD$ is quoted. Extended aeration and oxidation ditch systems may consume $125 m^3 O_2/kg BOD$ removed.

Operating problems with activated sludge

The most common operating problems with activated sludge plant largely revolve round the physical characteristics of the sludge, itself reflected by the biomass composition. Microscopic examination is useful here. The sludge flocs should be compact and regularly shaped with a few filamentous bacteria, and free moving animal life. Such a sludge will settle well, provide a clear supernatant and have a stirred sludge volume index (SSVI) of 100 ml/g or less. This test, described more fully in Chapter 2, is simple to perform with inexpensive equipment, and should be carried out daily.

(1) Rising sludge in the settlement tanks is a result of denitrification and occurs readily above ambient temperatures of 20°C. The main causes are low MLSS levels, a low RAS recycle rate or excessive aeration with DO levels in the plant of 4 mg/l or more. Collectively, this often points to poor control of aerators and pumps and inadequate DO measurement. The filamentous organism Microthrix parvicella is often present to com-

pound the settlement problem. Reducing the flow rate to the tank or increasing the sludge withdrawal rate usually effects a cure.

(2) The sludge settles poorly and much is lost over the effluent weirs. This is called 'bulking'; SSVI values are over 150 ml/g and the sludge blanket in the settlement tank is high up and not compacted. Microscopically, the sludge may appear thin, with few protozoa present. Alternatively, the sludge may be high in bacterial filaments, contain much bound water and be of low density, thus preventing compaction of the floc-forming bacteria.

Causes are insufficient DO in the aeration plant or a low F/M ratio. Nutrient imbalance, for instance high carbohydrate, low N or P, or shock loads which radically alter the carbon/nitrogen ratio can have the same effect.

Chlorination can be used to control the filamentous organisms by dosing the RAS stream with 1–3 mg/l free chlorine, but the effects of persistent chloro-organic compound formation and release to the environment should be considered. Often, altering the operating parameters will provide the cure.

(3) A stable foam forms on the aeration and settlement tank surfaces. The floc may not appear filamentous microscopically, but due to surfactants or long-chain fatty acids, large numbers of filaments may collect at the air/sludge interface. The predominant organism is *Nocardia* spp. Low sludge recycle rates, high DO levels and warm conditions often induce this state, but foam on the surface of oxidation ditches in particular is quite normal, if slightly unsightly.

In the author's experience this does not affect effluent quality providing that it stays in the aeration tank and does not break through to the final settlement tanks. This will depend on the flow variation within the system and the discharge flow rate from the aeration tank over the weir.

If foam does get through to the final settlement tanks, it will contribute to the suspended solids in the works effluent and consent failure is likely. Surface skimming pumps, also used to capture screenings that eluded preliminary treatment (mentioned in Chapter 1 and illustrated in Figure 1.2) have been successfully used to catch surface foam.

Because of their importance to successful operation, the process control parameters of activated sludge plants are extensively documented in the literature. Foot (1992) provides some guidelines as outlined above, but there are many other authors, and readers are advised to consult the journals of the relevant bodies concerned with wastewater management in their own country.

Anoxic zones

The benefits of anoxic zones in activated sludge plants were explored in the 1960s and may be summarised in the following ways:

(1) A power saving by reducing the number of aerators.

(2) Reducing the nitrate levels in the effluent. This decreases the chances of denitrification in the settlement tanks and often aids sludge settling characteristics as filamentous organisms are disadvantaged at very low oxygen levels and high organic loadings that characterise anoxic areas. Effluent low in nitrate also assists watercourses prone to becoming eutrophic by reducing nutrient input. Although rarely imposed at the present time, nitrate standards of 5–10 mg/l may be set in the future for plants in sensitive areas.

An anoxic zone is defined as one where no free oxygen is available, ie. the measurable DO level in the liquid is zero, and thus respiring organisms are forced to utilise the nitrate present as an oxygen source. They are normally generated at the influent end of the plant, where RAS and settled sewage or industrial effluent are mixed but not aerated. The high oxygen demand of actively respiring RAS and the BOD demand of settled influent quickly consume any free oxygen.

In the lack of a separate, small tank of short retention time, the first lane, or part of it, in the aeration plant can be made anoxic. To avoid settlement, submerged propeller stirrers are used to keep solids in suspension and assist the forward flow of the liquid. Typically, a 50% reduction in effluent nitrate can be achieved. Adequate DO monitoring is essential, coupled to mixer control.

Anoxic zones are used to control filamentous organism growth and thus bulking sludge successfully, and recent work (Foot *et al.* (1993)) indicates that careful design of the 'selectors' (small volume, short retention tanks) at the influent end can assist with controlling surface foam.

Some of the earliest work was done at Rye Meads STW, UK, by simply removing and blanking off half of the diffused air domes in the first lane of the aeration plant. Subsequently, anoxic zones were generated in other lanes experimentally although most contemporary activated sludge plants generate one zone at the influent end only. The operating parameters for extended aeration are, however, deliberately designed to generate anoxic conditions at several points in the aeration tank and are therefore quite different, as described in the next section.

Shock loads and toxic discharges

Activated sludge, like any other biological water treatment system, operates best under stable conditions, but even settled sewage varies in strength by a factor of 2:1 over a 24-hour cycle, and in industrial environments, shock loads are a common event.

Organic load variations usually present the problem of fluctuating oxygen demand and DO levels in the tank and varying sludge settling characteristics. Changes to nutrient ratios are more difficult to monitor continuously, but are equally likely to affect DO levels and favour rapid growth of particular organisms. Sludge bulking, caused by proliferation of the organism *Sphaerotilus natans*, can occur very rapidly at a sewage works receiving, for example, a sudden discharge of carbohydrate waste from a sweet factory.

Although some adaptability may be bred into the biomass, toxic organics such as chlorinated hydrocarbons, and toxic metals are likely to have a devastating effect on satisfactory operation. The degree of effect will depend on the mass of toxic material in relation to the total activated sludge plant volume. Common toxic metal levels above 5 mg/l in the aeration plant are likely to exceed the buffering capacity and any inherent biological resilience. A rapid deterioration of both effluent quality and sludge settling characteristics is very likely. Chlorinated hydrocarbons produce pronounced effects above 0.1 mg/l. In this respect, activated sludge is probably less hardy than the biofilter, where biomass turnover and metabolic rates are slower and more influenced by ambient temperatures, and biofilters often survive toxic discharges better than activated sludge.

Solutions to these problems include adequate numbers of strategically placed electrodes to monitor DO levels and a rapid response control system to alter aeration levels where organic load variation is likely. Nutrient ratio variations and toxic materials are best coped with by an upstream buffering or holding tank, apart from and preceding primary settlement and which is frequently monitored for likely toxics and/or nitrogen/phosphorus values as appropriate. Adjustment for the latter can then be made by dosing urea or phosphoric acid respectively.

The toxics are ideally controlled at source, either in the factory production area or through an effective trade effluent control policy of liaison and frequent visits and sampling, particularly to those sites known to use chemical toxic to activated sludge treatment.

In either case, the presence of a balancing tank gives some breathing space in which to analyse the contents, and then divert or bleed the discharge slowly to the plant as necessary.

The treatment of very variable flows, both in volume and strength, at small establishments such as caravan sites, military camps and trade effluent batch processes is probably best done by one of the 'fill and draw' systems. Here, the activated sludge tank is aerated for up to 24 hours, settled for 1–2 hours, and the top liquor or effluent discharged over a lowered weir or through floating arm valves. Biwater Treatment Ltd is one of a number of manufacturers who supply such a system, whose operation is more akin to extended aeration plant which are described next.

4.5.5 Extended aeration (oxidation ditches)

Extended aeration is a development of the original activated sludge system with several fundamental differences.

(1) The aeration tank is fed with crude sewage or industrial waste water, i.e.

no primary settlement. It is, however, essential that preliminary treatment to remove rags, grit and detritus is efficient. Reference has already been made in Chapter 1 to the decline in effluent quality attributable to a 25% loss in aeration tank treatment volume, itself entirely due to grit deposition.
(2) MLSS levels are maintained at a higher level of 4000–5000 mg/l.
(3) Retention time in the tank is typically 24 hours, or several days in some small, agricultural and industrial applications. Some small plant operate by 'fill and draw'.

(4) Aeration tank capacity is consequently larger than for activated sludge; 7000 m^3 is a common sized unit in the UK water industry. Oxidation ditches are thus able to tolerate shock and toxic discharges better, by having more dilution and buffering capacity.
(5) DO levels are lower, typically 15% saturation (roughly 1 mg/l at 20°C) at the aeration point. Aeration is exclusively by mechanical surface rotors, usually of the horizontal brush variety. Developments using aeration mats are currently in progress. The control of DO by electrodes placed near the aerators is by lowering the effluent weir. The top water level and rotor immersion depth are thus altered. Other control loops raise and lower aerator gearboxes, or turn aerators on and off. The first option of weir height control is the most common.
(6) Anoxic zones are a design feature of ditches, over 75% of the tank being anoxic without detriment to effluent quality. Thus, production of a denitrified effluent is a central feature, and typical effluent quality very high. BODs, suspended solids and nitrate values of less than 5 mg/l, and an ammonia of less than 1 mg/l are usual performance figures.
(7) The sludge tends to settle well because the large anoxic zones discourage filamentous organisms and because of the low nitrate in the effluent, denitrification in settlement tanks should not occur. It can, therefore, be retained for longer and a sludge blanket of roughly 15 000 mg/l solids easily generated. This is highly desirable as a preliminary sludge-thickening stage.

Operating problems

One of the disadvantages of oxidation ditches is that they produce a lot of sludge, typically 50%–60% more than accounted for by the incoming solids load; 1 kg/kg BOD removed is often quoted from operational plant. This can lead to an unhappy state of affairs (in the absence of sufficient sludge storage and dewatering facilities) in which MLSS levels have to be allowed to rise in the ditch and nitrification is lost, the ditch becoming nothing more than an aerated sludge lagoon. The material is also difficult to dewater above 2.5% by mechanical means, although polyelectrolytes are useful here. Co-thickening with raw sludges is not desirable as considerable layering is likely as the high soluble nitrate content is released as nitrogen gas.

For the same reason, any co-settlement with raw sludges in primary settle-

102 *Sewage and Industrial Effluent Treatment*

ment tanks (in cases where the oxidation ditch runs in parallel with a conventional biofilter plant) is likely to induce rapid nitrogen release and disastrous tank inversions. The advice is to keep oxidation ditch sludge separate.

Another minor problem is the often prolific growth of a thick surface foam or scum on most of the aeration tank surface. The low DO levels in the anoxic zones favour development of *Nocardia* spp., which is unsightly, but this is often not a treatment problem providing it does not break through to the settlement tanks. The aerators will tend to break up this material as it passes through, and this can be induced by periodically raising the top water level to increase circulation rate. Other methods, such as silicon oils, are not worth the expense and are only temporary in effect.

Figure 4.10 shows an oxidation ditch (one of two side by side) of $6800\,m^3$ each, treating domestic crude sewage, which enters on the bend, along with returned activated sludge (RAS). The effluent weir is to one side before the first aerator. Note that one lane is not aerated, and thus the ditch is anoxic for the majority of the outside channel and about 50% of the two inside channels. Typically run at 5000 mg/l MLSS to produce an effluent of the standard mentioned above, this plant has experienced operational difficulties with sludge disposal in winter, and a lack of storage capacity has led the plant to run at 8000 mg/l MLSS for short periods.

Above 6000 mg/l, nitrification is progressively lost at this site and only carbonaceous oxidation achieved, resulting in high effluent ammonia values. The tanks are 3.5 m deep, and only 25% buried, saving on excavation and

Figure 4.10 Oxidation ditch.

construction costs. DO measurements taken by one electrode control effluent weir height and rotor immersion depth.

Average retention time is about 22 hours and cycle time 17 minutes. The plant produces typically 45% surplus sludge, which after 30 days' storage and dewatering is about 3% DM. After 11 years of operation, it is becoming apparent that the ditch contains a lot of grit that eluded the preliminary trap and will require draining and excavation shortly.

Extended aeration design parameters

Oxidation ditches were first developed in 1953 by Pasveer to simultaneously treat sewage and aerobically digest sludge in one tank. As no primary sedimentation is usually involved, the loading rates are less than those for activated sludge.

(1) The organic loading is set at 0.05 kg BOD/kg MLSS/day.
(2) The volumetric loading is set at about 0.2 kg BOD/m^3/day.
(3) The retention time of the solids or sludge age is often 15–25 days.
(4) Installed aeration capacity is frequently about 0.5 kg O$_2$/m^3/day, about half that for activated sludge, and underlines the reliance on anoxic conditions to induce oxgen recovery from free nitrate. This low level of aeration also precludes aerobic sludge digestion. It has been estimated that 3 kWh is required to destroy 1 kg of sludge, and operation in this mode would therefore be uneconomic at sewage treatment sites.

Some interesting work has been carried out to evaluate the oxidation ditch for treating sewage and average strength organic waste (Carmichael & Johnstone, 1982; Johnstone & Carmichael, 1982), where the 'package' nature, cheap construction and high quality denitrified effluent have led to their ready acceptance worldwide.

4.6 Conclusions

This chapter has attempted to give an overview of operating aspects of biological oxidation. Readers are advised to refer to the considerable data available by computer retrieval or the journals of relevant learned societies from which many papers of practical operating experiences can be unearthed.

Due to the very adaptability of biological systems, only design ranges are appropriate, and some of the results in the literature may appear semi-contradictory. The best advice is therefore to design and initially operate the plant within accepted ranges, and then experiment, giving the system time to adapt to the new regime over a period of at least one month. This will allow at least one 'sludge age' to have elapsed.

There are contemporary variations on all biotreatment systems offered by many manufacturers and suppliers. 'New' developments are constantly appearing, including aerated and anaerobic flooded filters and Deep Shaft treatment which, along with others are oulined in Chapter 6.

4.7 Summary table of typical operating parameters and dimensions

4.7.1 Requirements common to all secondary biological systems

Optimum pH range	7.0–8.5
Nutrient balance ratio of BOD:N:P	100:5:1
Limiting elements	K, S, Ca, Mg, Na and Fe
Micro-nutrients	Zn, Mn, Mo, Se, Co, Cu, Ni, V and W

4.7.2 Biofilters

Feed	settled sewage
Depth	2 m average
Circular filters	4–20 m dia
Media	size range 25–150 mm dia
Specific surface area 28 mm slag	220 m^2/m^3 media
Typical voidage	40%–80%
Liquid retention time	20–35 min for 80%–90% reduction
BOD percentage removal	60%–90%
NH$_3$N percentage removal	50%–90%
Bacterial removal	70+%
Hydraulic loading rate	0.60 m^3/m^2/day for 90% nitrification
	0.80 m^3/m^2/day as an upper limit for 80% removal of BOD and ammonia
Biological loading rate	0.1 kg BOD/m^3/day for full nitrification

(The British Royal Commission Standard gives 0.12 kg BOD/m^3/day to effect the highest BOD removal and nitrification.) (The boundary between low and high rate filtration is represented by 0.2 kg BOD/m^3/day for 80% BOD removal. A poor ammonia removal of less than 50% may be expected at this rate.)

Ammonia loading rates	0.05–0.08 kg NH$_3$N/m^3 day (conventional hydraulic rates)
	0.08 kg NH$_3$N/m^3/day for 60% removal

4.7.3 High-rate biofilters

Feed	settled sewage
Depth	3–7 m
Media	randomly packed plastic/corrugated sheets
Specific surface area	300+ m²/m³ depending on media shape
Typical voidage	90%
Liquid retention time	less than 10 min
BOD percentage removal	30%–70%
NH$_3$N percentage removal	10%–50%
Bacterial removal	40+%
Hydraulic loading rate	5–10 m³/m²/day is common 25 m³/m²/day is possible
Biological loading rate	1 kg+/m³/day
Ammonia loading rate	0.08 kg NH$_3$N/m³/day for 60% removal

4.7.4 Rotating biological contactors

Feed	settled sewage
Media	close-packed corrugated plastic discs, 40% immersed
Surface area	9000–16 000 m² for corrugated high-density polypropylene discs on an 8 m long shaft
Liquid retention time	0.5–4 h
BOD percentage removal	50%–80%
NH$_3$N percentage removal	10%–60%
Bacterial removal	40+%
Hydraulic loading rate	0.08–0.16 m³/m² of disc surface area/day
Biological loading rate	0.01 kg soluble BOD/day for 60% BOD removal

4.7.5 Activated sludge

Feed	settled sewage
Tank length/width ratio	4:1–10:1

(Some plants consist of a series of lanes round which the liquid flows, each lane being 30–50 times longer than its width.)

Tank depth	3 m average
Liquid retention time	6–8 h
BOD percentage removal	90+%
NH$_3$N percentage removal	90+%
Bacterial removal (all in nitrification mode)	90+%
Mixed liquor suspended solids (MLSS)	2200–3000 mg/l for reliable nitrification
Volumetric loading rate	0.5–1.0 kg BOD/m^3/day
Organic loading rate (F/M) ratio	0.1–0.25 kg BOD/kg MLSS/day for full nitrification 0.4–0.6 kg BOD/kg MLSS/day for carbonaceous oxidation only
Maintained DO level	1.5–2.0 mg/l
Kg oxygen supplied/kWh	2 kg O$_2$/kWh (1.3–1.6 are more typical values)
Installed aeration capacity	1.0–3.5 kg O$_2$/m^3/day
Oxygen availability	2 kg O$_2$/kg MLSS/day for full nitrification
Oxygen requirement	
for carbonaceous oxidation	1.0–1.2 kg O$_2$/kg BOD removed
for nitrification	1.6+ kg O$_2$/kg BOD removed
Oxygen requirements – F/M ratio greater than 0.3	30–55 m^3 O$_2$/kg BOD removed for coarse bubble diffused air 25–35 m^3 O$_2$/kg BOD removed for fine bubble
Oxygen requirements – F/M ratio less than 0.3	75 kg m^3 O$_2$/kg BOD removed where nitrification must be guaranteed
Sludge age	10–20 days
Sludge SSVI	40–150 ml/g
Sludge production rate	0.5–1.0 kg/kg BOD removed
Recycle/RAS return ratio	1:1–2:1

4.7.6 Oxidation ditches and extended aeration systems

Feed	screened crude sewage
Tank depth	1.5–3.5 m
Liquid retention time	24+ h
BOD percentage removal	90+%
NH$_3$N percentage removal	90+%

Bacterial removal (all in nitrification mode)	90+%
Mixed liquor suspended solids (MLSS)	4000–5000 mg/l
Volumetric loading rate	0.2 kg BOD/m^3/day
Organic loading rate	0.05–0.15 kg BOD/kg MLSS/day
Maintained DO level	0.7–1.5 mg/l
Kg oxygen supplied/kWh	2 kg O$_2$/kWh (1.3–1.6 are more typical values)
Installed aeration capacity	0.5 kg O$_2$/m^3/day
Oxygen requirements – F/M ratio less than 0.3	125 kg m^3 O$_2$/kg BOD
Sludge age	15–25 days
Sludge SSVI	40–150 ml/g
Sludge production rate	1.0 kg/kg BOD removed
Recycle/RAS return ratio	1:1–2:1

4.8 References

Carmichael, W.F. & Johnstone, D.W.M. (1982) Initial operating experiences at Cirencester Carrousel Plant. *Journal of the Institution of Water Pollution Control*, **81** (5).

Foot, R.J. (1992) The effects of process control parameters on the composition and stability of activated sludge. *Journal of the Institution of Water and Environmental Management*, **6** (2).

Foot, R.J., Robinson, M.S. & Forster, C.F. (1993) Operational aspects of three 'selectors' in relation to aeration tank ecology and stable foam formation. *Journal of the Institution of Water and Environmental Management*, **7** (3).

Hemmings, M.L. & Wheatley, D. (1979) Low rate biofiltration systems using random plastic media. *Journal of the Institution of Water Pollution Control*, **78** (1).

Hill B., Toddington, R. & Challenger, P. (1992) Burnley STW – design and operation of a high-rate nitrifying filter. *Journal of the Institution of Water and Environmental Management*, **6** (3).

Johnstone, D.W.M. & Carmichael, W.F. (1982) Cirencester Carrousel Plant: some process considerations. *Journal of the Institution of Water Pollution Control*, **81** (5).

Thomas, C. & Slaughter, R. (1992) Phosphate reduction in sewage effluents: some practical experiences. *Journal of the Institution of Water and Environmental Management*, **6** (2).

4.9 Companies and other organisations

Biwater Treatment Ltd, Biwater Place, Gregge Street, Heywood, Lancashire OL10 2DX.
BS Floccor Ltd, E5 Stafford Park 15, Telford, Shropshire TF3 3BB.
ELE International Ltd, Eastman Way, Hemel Hempstead, Hertfordshire HP2 7HB.

Klargester Environmental Engineering Ltd, College Road, Aston Clinton, Aylesbury, Buckinghamshire HP22 5EW.
Mass Transfer International, Heversham, Cumbria.
Rosewater Engineering, Stonebroom Industrial Estate, Stonebroom, Derbyshire DE5 6LQ.
Warren Jones Engineering, 120 Churchill Road, Bicester, Oxfordshire OX6 7XD.

Chapter 5
Sludge Disposal and Treatment

5.1 Introduction

This chapter comprises two main sections:

(1) Disposal strategies, and the adoption of as positive an approach as possible to a material that is frequently regarded as a nuisance, and often an expensive one.
(2) Contemporary treatment methods, where the emphasis is on maximising the benefits derived from simpler techniques to reduce volume and disposal costs.

With very few exceptions, all the treatment systems described in earlier chapters produce or generate sludge, or several different types of sludge in multi-stage processes, which often have different characteristics and may require separate treatment.

Apart from the deliberate generation of sludge by an initial settlement stage to reduce the suspended solids (SS) content of the wastewater, or chemical precipitation to remove unwanted or toxic components, all biological processes generate sludge as a result of oxidation and the cyclic regeneration and death of the biomass effecting treatment. Settlement after biological treatment is essential to remove the solids generated and produce a high quality effluent, as well as to retain and recycle within the plant the living biomass of solids to maintain the biological process. It is important to appreciate the volume of sludge some of these systems can produce as, regrettably, the plant that 'consumes its own sludge' is normally confined to manufacturers' sales literature.

Some biological treatment systems would appear to produce, in normal operation, over twice the mass of solids that can be accounted for in the solids load originally entering the plant. It should be mentioned here, however, that not all the solids entering the treatment plant are necessarily organic and biodegradable. Up to 50% of crude sewage solids may be inorganic, and there is a continuous loss of solids in the effluent of all plant which is often not accounted for in mass balances.

Certainly, if a solids mass balance through a plant is required, a considerable number of flow-weighted bulked samples are needed to account for all the additions and losses. As a particular example of the biological generation of sludge, oxidation ditches, devoid of a primary settlement stage, are frequently reported as major sludge producers, and values as wide as 0.2−1.3 kg sludge/kg BOD are reported in the literature (see Carmichael & Johnstone, 1982).

Such extended aeration systems are, however, often operating as aerobic sludge digestors, a considerable volume of the oxygen supplied being used to destroy biomass; 3 kg O_2/kg sludge destroyed is a reasonable working figure for this form of sludge 'treatment' − an expensive method of sludge disposal unless energy is virtually free. It is equally important to appreciate the different characteristics of the various sludges produced, and contemporary methods for treatment to a disposable or recoverable form are outlined in subsequent sections.

Treatment methods can be very simple, ranging from dewatering under quiescent conditions in order to reduce the volume of sludge through to the high complexity of total destruction of all organic matter by incineration − leaving only an inorganic ash for disposal.

In all cases, some from of dewatering or drying is carried out first. The water in most inorganic sludges is weakly bound within the solids matrix, and gravity settlement usually produces good separation and a well-compacted sludge. The water content of many organic sludges, however, is quite firmly bound to organic molecules by a mixture of physical and electrostatic forces. While gravity settlement will achieve some separation, the success of many sludge treatment systems relies on correct dosing with chemicals. This will reduce the attractive charge between water and solid particles, and facilitate removal of much more water. A pH change may also be beneficial. The process is often assisted by applying mechanical pressure to squeeze out the water.

It should not be overlooked that organic sludges are a considerable potential energy source. Providing no inhibiting materials (chlorinated solvents, for instance) are present, sewage sludges, those of farm origin and many industrial cocktails will anaerobically digest to produce methane/carbon dioxide in roughly 70/30 proportions. This gas has a calorific value equal to, and often in excess of, domestic supplies. Apart from heating, it is widely used in the water industry to power dual fuel or dedicated gas engines driving alternators, and large sewage works are frequently self-sufficient and energy exporters of either bottled gas, heat or electricity. Digestion times are typically 15−30 days at 37–40°C. The end product is much less odorous, more stable and still retains its fertiliser value as nitrogen and phosphorus components. A volume reduction of about 20% occurs.

The method requires some expert supervision, and an appreciation of the hazards associated with methane, but it has much to recommend it, as witnessed by the development of small, automated plant designed specifically for farming applications.

5.2 Initial considerations

There are some basic points to note when selecting a particular wastewater treatment system and with regard to the treatment of the sludges produced.

(1) Look carefully at the type and volume of sludge a new treatment system is going to produce, based on previous, documented applications and examples from within a similar industry. This is particularly important if transportation to the dwindling number of landfill sites available in many countries is the preferred sludge disposal route.
(2) Generously size the sludge handling and storage area. Allow for wet weather, sludge treatment plant breakdown and transport difficulties.
(3) Identify a long-term, stable disposal route with predicted costings.
(4) Minimise sludge production by good housekeeping in the factory and subsequently stored and transported volumes, and recycle wherever possible. Isolate high solids discharges and dispose of or treat separately.
(5) Aim to remove as much water as possible from sludges before treatment or disposal. Remember, however, that pumping sludge that is more than about 10% dry matter is difficult and expensive.
(6) View the material positively; if separated or treated in a particular way, it may have further economic uses on site as, for instance, a fuel, feed supplement to livestock, fertiliser, process additive or backfill material. Alternatively, its contents may be recoverable and saleable within your type of industry or a neighbour, to mutual financial advantage.

5.3 Sludge types and characteristics

5.3.1 Inorganic sludges

Many inorganic sludges represent waste of raw materials from separation processes, and the materials should always be recovered where possible and blended with feedstocks. The basic properties are those of inertness, and a tendency to settle quickly and form dense deposits. Inorganic sludges vary in density between about 1.5 and 2.5. Sludges and slurries from mining coal, clays, minerals and fine particulates such as chalk, sand or iron and steel dusts are in this category and will settle rapidly, often compacting and requiring mixing or slurry formation to move or pump. A lagoon of these materials usually requires digging out to remove successfully. In view of their unreactive nature, odour problems arise only from organic contamination or bacterial action reducing sulphates to sulphides.

Fibrous materials of plant origin, paper or hair will also form dense sludges fairly easily without too much attention having to be paid to the design of a simple settlement tank or lagoon. Metal hydroxide sludges, formed by reaction with sodium hydroxide or lime to remove toxic elements from wastewaters as required by discharge Consents, are light and flocculant, and require quiescent settlement conditions in a tank with inlet and outlet baffles, a sloping floor and regular de-sludging regime (see Chapter 2).

5.3.2 Organic sludges

Organic sludges are frequently odorous and likely to undergo biological breakdown with the release of noxious/flammable/explosive gases like methane, hydrogen sulphide and hydrogen, along with nitrogen and carbon dioxide. Such reactions are temperature dependent. As the water content is often biologically bound, dewatering is more difficult, and in quiescent conditions, bands of water and sludge are quite common during settlement. The density of most sewage works sludges lies between 1.03 and 1.06. Untreated 'raw' organic sludges represent a potential source of energy by digestion or incineration and sewage sludges are a valuable soil conditioner and fertiliser.

After digestion, the sludge will have an earthy odour and be far more pleasant to handle. The organic content will have dropped by about 25% and the sludge will be more difficult to dewater. Agricultural disposal is less restrictive for digested sludges in most countries as the process also reduces the number of viable pathogens. The nutrient composition of sewage-based sludges is shown in Table 5.1.

Table 5.1 Nutrient composition of sewage-based sludges.

Content	Primary (%)	Secondary (%)	Activated (%)
Grease and oil	6–30	5–20	5–12
P as P_2O_5 on DM	0.8–2.8	1.5–4.0	2.8–11
Total nitrogen on DM	2–3	3–4	4–8
K as K_2O on DM	>1	>3	0.5–0.7

5.4 An overview of treatment options

Recent organic sludge treatment and disposal practices in Europe have become largely polarised towards the high-tech solution of incineration or a revamping of some old ideas including composting, where the material is recycled to land with or without digestion. The incinerator can become, with the right feedstock, a net exporter of heat and/or electricity, and some ambitious schemes based on sewage sludge are under construction. For large organisations, with a lot of sludge on one site and which cannot readily be transported or converted, incineration is attractive, and particularly where other disposal routes are being denied.

The UK water industry has been forced down this road at a number of coastal sites, following the banning of sewage sludge disposal to sea after 1998 (see North Sea Ministerial Conference, 1991). The justification for this political gesture on environmental grounds, when taking all the factors into account, is doubtful, and the long-term outcome may prove expensive.

At the other end of the technical scale, forced air or covered composting with straw has seen a revival. If the organic sludge is uncontaminated, land disposal is local and plentiful, and your site has plenty of space and tolerant neighbours this is cheap and environmentally sound.

Somewhere in between these extremes lie a number of possibilities which will suit the majority of industrial sludge producers where expertise and space are not plentiful and the following are likely to be more attractive options.

5.4.1 Dewatering by quiescent settlement for 1–7 days (depending on the sludge)

This is very simple, requires only one or two tanks, a few valves and a couple of minutes of time per day. And yet as a method for reducing the volume of sludge for disposal, it is often neglected to great financial loss. Most sludges will dewater well, and reduce the sludge volume by 50%–70%.

Mechanical dewatering or drying to produce a cake or pellets is fairly simple and will also substantially reduce volume. In this form, sludges from the food industry can be high in protein and a valuable animal feed supplement. Fibrous inorganic sludges based on paper products tend to dewater very readily, and are relatively innocuous for landfill but they may have sufficient calorific value when formed into brickettes to warrant use as a supplementary boiler fuel. Oil- and paint-based sludges may have similar calorific potential, although their burning may cause failure of tight stack emission standards unless carefully controlled or mixed with other fuel. Solvent and oil recovery first are often better options here.

Arguably, dewatering is one of the best techniques for industry. Equipment is not very expensive, successful operation requires only basic training, and there are often tenfold volume reductions or more during mechanical dewatering.

If off-site disposal is necessary and the product is literally waste, skips to take sludge cake away are presently about half the price of a tanker to take away the equivalent liquid volume.

5.4.2 Fixing metal content in an insoluble matrix

Any inorganic sludges containing metals are best concentrated, and in view of the problems of leaching in landfill sites and strict disposal conditions imposed in many countries, the best option is to extract or at least fix the metal content in an insoluble matrix. If oily, or contaminated with organics, incineration followed by metal recovery is another, increasingly important option as the prices of raw materials rise.

5.4.3 Digestion

Digestion has a number of attractive aspects for organic sludges of sewage or farm waste origin, and those from the brewing and food processing industries.

Cold digestion

Cold digestion at ambient temperature will proceed slowly in cool climates, and will stabilise, dewater and reduce the volume of sludge, but this requires land and can be odorous. Nevertheless, an industry producing erratic and small quantities of organic sludge may find a sludge lagoon where the retention time is measured in years, a useful 'bottomless pit', organic breakdown and dewatering mostly counterbalancing addition of fresh sludge. Some mechanical mixing and top water withdrawal is often beneficial, particularly if annual rainfall is high.

Forced aerobic digestion

Forced aerobic digestion where air is blown into the sludge to accelerate the breakdown of solids is energy-intensive, but this would be practical in warm/hot climates where electricity is cheap, and odorous components stripped out would not cause offence.

Heated mesophilic digestion

Heated mesophilic digestion at temperatures of between 35°C and 45°C is normally carried out anaerobically in closed tanks for 15–30 days. Apart from the bonuses of a low-odour, stable product, with a reduction in volume of 20%–30%, the process yields digestor gas – typically 70% methane and about 30% carbon dioxide with traces of sulphides and mercaptans. The calorific value of this gas equals that of most municipal gas supplies and represents a valuable energy source. After heating the digestor, the surplus can be used to heat buildings, and on large sites generates electricity from dual-fuel diesel/gas engine/alternator sets.

Most large sewage works sites use this system of sludge treatment, and are usually energy self-sufficient; some sell any surplus gas or compress it to drive vehicles, while a few contribute to local heating schemes. Digestor gas gives a clean burn, and the only difficulties experienced in gas engines are associated with trace levels of hydrogen sulphide converting to sulphuric acid and causing internal corrosion. A simple gas scrubber provides a cure.

Anaerobic digestion needs careful management and monitoring, however, and benefits from economies of scale, making it inappropriate for small-scale industrial situations where the average daily sludge volume is less than about $1 \, m^3$. Methane/air mixtures are explosive and planning and safety restrictions tend to confine its use to open sites away from housing, roads and bridges and adjacent factories. Industrialists contemplating this method of sludge treatment must ensure that on-site expertise is regularly available.

5.4.4 Treatment options for sewage sludge

In sewage treatment there is little control over the raw material other than that imposed by trade effluent control, and sludge volumes and compositions markedly differ between works.

Inadequate sludge storage, handling, dewatering and disposal facilities hamper works operation, and many a sewage works manager would describe sludge as the '*bête noire*' of the whole process. Over-loaded facilities are often a source of unpleasant odours, to which the public are increasingly sensitive. The trend to energy-intensive aerated systems producing a lot of sludge that is difficult to dewater has not improved matters.

The Urban Wastewater Directive (91/271/EEC) (1991a) requires much higher sewage treatment standards, and in the UK particularly the treatment of untreated sewage discharges to sea.

It has been predicted that sewage sludge production in the UK will double by 2006 from the present 1.1 million t dry weight currently produced per year. As has been mentioned, from December 1998, the UK will also lose the sea disposal route, by which an estimated 29% of sludge travels – about 0.43 million t DM/annum. These two factors are causing some uncertainty as to future sludge disposal routes and the respective proportions that will go to land or be incinerated.

It is likely that beneficial recycling of sludge to agriculture will remain the biggest single outlet. An estimated 465 000 t dry solids is so disposed in the UK at present, a volume which might double by 2006. Other land-based and beneficial routes capable of expansion include forestry and land reclamation uses such as landscaping and capping of landfill sites, derelict land rehabilitation and as a topsoil substitute on major roadworks.

These outlets are regulated by the Sludge (Use in Agriculture) Regulations 1989, which implement the EC Directive 86/278/EEC (1986). The quality of the sludge will dictate suitability here. The removal or absence of non-biodegradable plastics is one particularly relevant factor in gaining acceptance.

There is certainly room for expansion for recycling to land, providing that effective trade effluent control of discharges containing the heavy toxic metals (Cu, Ni, Pb, Cd, Cr, Zn) is stringently applied by the Water PLCs to avoid excessive accumulation on land, diminishing outlets and decreasing application rates.

Prospective EC Directives that may affect all land disposal routes are the Hazardous Waste Directive (91/689/EEC) (1991c) – it is unlikely that sludge will be considered hazardous in the waste context – and the Nitrates Directive (91/676/EEC) (1991b). The latter may restrict sludge application rates and the timing of land applications to minimise nitrate leaching to ground and surface waters.

It is clear that incineration is becoming a favoured option and especially for sludge from coastal sites or those on estuaries where well-established arrangements for shipping the material seawards have been in place for a long time. The size of some sites and the sludge volumes have precluded the use of road tankering to land on environmental impact grounds and it is predicted that 20% of UK sludge will be incinerated by 2006.

The 17th Report of the Royal Commission on Environmental Pollution views incineration as a beneficial outlet of low environmental impact in view of the auto and often exothermic operation now possible and the stringent gas emission standards now applied. The general public to date has been less enthusiastic, adopting the much-publicised 'Not In My Back Yard' (NIMBY) stance and influencing planning approval.

If the EC Directive 91/689/EEC (1991c) does classify sludge as a toxic material, sludge incineration will become more fraught, as the ash will be similarly classed.

It is important to remember that as sewage sludge is typically 75%−80% organic, 20%−25% of the original sludge volume will remain as ash for landfill disposal after incineration, an ash containing by concentration 3−4 times the original toxic metal content. This might yield a material containing mixed metals of 1%−2% by weight, where the original toxic metal content of the sludge is 400−500 mg/kg DM, and similar in metallic content to some mining ore. Although the leaching of metals from the ash is unlikely in normal ground conditions, co-disposal with domestic waste in landfill sites generating volatile acids is likely to create metal-laden leachate.

It is particularly important to have confidence in the energy output of incinerators, for the operating economics of some present designs rely heavily on achieving at least autothermic operation. To achieve autothermic operation, a sludge DM in excess of 28% is required from considerable operating experience in the UK and USA. Exothermic operation relies on consistent cake quality in excess of about 33% DM and calorific values above 20 MJ/kg DS. The use of organic based polyelectrolytes for sludge dewatering will boost this value.

Marginal changes in calorific value and DM caused by variability in sludge properties will have considerable impact on the amount of excess energy produced. In the author's opinion it is unwise, therefore, to abandon other energy sources such as digestor gas and incinerate all the sludge in raw form, at large-scale undertakings. Incineration cannot be 100% reliable and there will be energy shortfalls at times requiring considerable energy to be imported to maintain the main works aeration processes and pumping. Very adequate sludge storage facilities to cover incinerator maintenance periods is also important.

Somewhere in between the extreme options of 'back to land' and 'high tech' lie others of dewatering, drying and digestion, and all can benefit from new technologies and the economic application of scarce resources. Some of the newer drying techniques produce a pelletised product of 70%−90% DM, easily handled and often marketable.

Sludge digestion yields a large energy source, and due to the development of combined heat and power (CHP) plant on smaller scales recently, offers automated power generation and possible sale of electricity to the grid under 'Green Tariff' incentives. Such developments are now practical for works of 20 000 population equivalent or more. Larger works are usually energy-independent; there are a number of shining examples in the world that produce excess digestor gas and sell it on to industry, and have done so for 60 years. 'Being green' isn't necessarily a recent phenomenon.

Digested sludge at 3% DM and more oxidised is more difficult to dewater than raw, but 25% is typically achieved by a belt press. Either in liquid form or as a cake, digested sludge has a low and reasonably accepted earthy odour, and finds ready outlets to land.

There remain many rural, unmanned sites where dewatering is the only treatment, the effectiveness of which dictates tankered volumes and costs. A survey of such sites reveals a wide range of facilities, and an equally wide variation in their effective use. The UK water industry currently spends 20% of its budget on sludge treatment and disposal, making it an area of operations that should receive an equally sizeable chunk of resources, good forward-looking design and continuous research.

There are too many examples, from the author's experience, of tanker loads of dirty water being moved around at considerable expense arising from a lack of operational control of simple dewatering facilities, and it is an area worthy of adequate resourcing, for one of the simplest and primary sludge strategies is to reduce the volume of material to the minimum. At a local level, and much dependent on circumstances, sludge disposal can swallow 50% of the operational budget.

Part of any ongoing sludge strategy is to assess risk relating to the various outlets. Factors such as long-term disposal site availability, reliability of transport, susceptibility to weather conditions, public safety and health hazards either remote or by direct handling must be considered.

Apart from present and certain future legislation and the likely outcome of proposed EC Directives, related issues such as a more stringent groundwater protection policy, diseases to farm animals, competition for sites within the water companies and shifts in public opinion must also be addressed.

A formal risk assessment is an appropriate exercise now being adopted by many water companies worldwide. If the system fails, what can go wrong, with what likelihood, what would happen and how can the risk be reduced?

In conclusion, and in view of the current uncertainty surrounding a number of sludge disposal techniques, now is an attractive time to explore new markets for sludge in which the material is 're-packaged' and the revamping of some old techniques such as composting where space allows. Above all, the potential energy of the material should be realised, particularly when the resulting product is more stable and easier to dispose of. This applies equally to digestion and incineration.

5.4.5 Treatment options for industrial sludges

Although the types and composition of industrial sludges vary widely, they tend, with some exceptions, to be less complex in content than sewage sludge; and another advantage is that the industrial operator has more control as to the ultimate sludge composition and volume, through good (or bad) work practices within the factory. The initial emphasis must always be on reduction of sludge volumes by simple techniques such as dewatering and chemical dosing, with exploration of all possibilities for recovery for reuse or sale. Where possible, isolating sludge production from different sections will facilitate recycling.

Conversely, if the sludge is organic and digestion is proposed, mixing to dilute toxic components may be necessary. In every case, a comprehensive survey of the range of sludge composition likely to be produced is required, for sludge digestion is a fragile process and time-consuming to re-establish. Nevertheless, it offers the industrial user with organic sludge and a daily sludge volume exceeding $1\,m^3$ a large potential energy source in the form of methane gas, if adequate expertise is available to handle this fuel and the attendant safety requirements.

If starting from scratch with a new treatment plant, critically examine the supplier's expectations (by reference to actual operating examples) of the sludge production of any proposed biological treatment system. Some of the new package systems produce very good effluent at some energy cost, as well as a lot of sludge which is difficult to dewater and may not digest readily, having been partly aerobically digested in the treatment plant. More than one enthusiastic operator of a new package plant has been heard to observe that the process is working very well, and then adds '... but it produces an awful lot of sludge, the neighbours complain about the smell, and we can't get rid of it...' (neither the sludge nor the smell).

Dewatering sludges by pressing to a cake of 25%–40% DM has much to offer; the process is reliable and less affected by quality and composition variations, the equipment robust, and the product often dry and more easily handled off-site by skip. Tenfold volume reductions may be achieved. If incineration is the ultimate step, many organically-based sludges above about 30% DM and 70%+ organic will burn autothermically.

If extraction is to be carried out, whether for metals or pharmaceuticals, the advantages of pre-concentration are obvious. The requirements here are to ensure adequate, correct chemical dosing rates for sludges that need this preliminary treatment to dewater easily by vacuum or pressure filtration, good mixing facilities and frequent cleaning of belts, cloths and plates to prevent blinding and maintain dewatering efficiency.

Incineration of industrial sludges requires on-site expertise to operate the equipment properly with regard to complete combustion at temperatures exceeding 850°C and to comply with demanding gas flue emission standards.

The latter have become much more demanding within Europe in recent times, and present standards are referred to in Section 5.7.4.

In the UK, incinerators, including those at sites originally 'Crown Exempt' such as hospitals, lie within new local authority air pollution control systems under Part 1 of the EPA 1990. But in less stringent cases, it remains important to avoid visual smoke production and ensure rapid high-temperature combustion to avoid dioxin and furan production (both highly toxic), and carcinogenic poly-aromatic compounds forming, e.g. benzo(a)pyrene.

Dust precipitators and gas scrubbing are thus normal features associated with sludge incineration from any source. Liquors from the latter may require biological treatment before discharge. The refractory nature of some organic condensates will require co-treatment of these compounds with other wastewater, or an intensive aerated treatment system.

In its favour, incineration offers total destruction of the organic fraction of any industrial sludge to a non-putrescible and sterile ash, the option of waste heat recovery and considerable mass and volume reductions. It is an attractive route for destroying organics resistant to biodegradation, and some useful concentration of metals may be achieved and their extraction made easier to assist recycling.

5.5 Treatment methods

5.5.1 Simple dewatering

The sludge is discharged into a tank or series of tanks and allowed to settle. Settlement characteristics will be entirely unique for each type, mixture or blend, but most sludges will settle to the bottom of the tank within 24 hours with a two to fourfold increase in sludge dry matter content. Typically, 1%−2% DM inflow will yield sludge of 4%−8% DM. Top water can then be decanted through a series of dewatering gate valves at different levels, or over a penstock valve or weir whose height can be adjusted. The former is illustrated in Figure 5.1, the tank being prefabricated grp-coated steel, and a common design in the water industry.

Sludge is withdrawn from the base for further treatment or transportation. Most processes, industrial and sewage, will make sporadic sludge/water discharges into a tank, which may prove a sufficient disturbance to hinder good settlement, or induce rising sludge where the organic content is beginning to denitrify/decompose. At least two tanks are therefore desirable, operated alternately. This will also be of benefit where the material forms bands of water and sludge, and some operator time is needed to open the dewatering valves in sequence to extract the maximum amount of water.

If insufficient room or resources are available for at least two tanks, ensure the discharge point is into the top of the tank or across the top water surface, and not into the base.

120   *Sewage and Industrial Effluent Treatment*

Figure 5.1   Grp-coated sludge dewatering tank and dewatering valves.

Slow stirring is beneficial for organic solids or fines. Figure 5.2 is a cross-section and illustrates the vertical bars of a picket fence widely used to thicken raw sewage sludge, reduce banding and release trapped gas which can prove a problem in hot weather by inducing solids to float. Biwater Europe Ltd is one of a number of UK suppliers of such equipment.

On unmanned sites, a practical way to effect at least some dewatering is to crack open most of the top valves on the tank(s), which will give automatic dewatering until the valve becomes totally blocked with solids. Providing the sludge solids are not too fibrous and the section of plant receiving the dewatering discharge will accept quite a high solids content at times, this 'trick of the trade' works very well with a little practice, and considerably reduces the time a visiting gang or operator has to spend dewatering by hand.

An external sight glass on the tank is practically of little use for dewatering

Figure 5.2 Cross-section through a picket fence sludge thickener/stirrer.

and usually quickly blocks; several manufacturers of grp or plastic tanks have inserted transparent plastic or glass window strips in the walls, and providing they can be kept clean they are beneficial in viewing the contents.

A sludge blanket detector, measuring optical transmission between a photocell and light source, will also detect sludge/water interfaces and finds application in settlement tanks to control sludge withdrawal rates (Chapter 2). They are less successful when trying to locate water bands in fibrous, finely divided, oily or raw sludges, as the light source becomes quickly fouled. ELE International Ltd supplies a convenient hand-held device.

Ultrasonic level detectors or float switches are widely used to determine top water level in holding tanks, the latter particularly as a high level alarm sensor. Both are very reliable and essential equipment for sludge storage tanks. Milltronics Ltd is one of several UK suppliers of ultrasonic detection systems, which by their non-contact with the fluid/sludge, do not suffer the fouling associated with immersed probes. Sarlin Ltd supplies an encapsulated float switch, widely used by the UK water industry to control pumps emptying wet wells, and to activate high level alarms via telemetry links.

Dewatering costs can be negligible, capital items apart, and are influenced primarily by the labour requirements which are often minimal. The capital costs of tanks is often quickly recovered by the considerable transport saving resulting from the volume reduction.

5.5.2 Composting

There has been a recent and interesting revival in this old technique. The main differences are that the modern method is more intensive. The material is mechanically turned at intervals, force aerated and often contained in a building where heat losses, odour and water content can be controlled.

The principal feature is the generating of heat by biological activity during decomposition of the material, which if retained will not only allow much faster degradation, but will also reduce pathogen levels. Success dictates that the compost must be maintained at 40°C for at least five days, achieving 55+°C in the main pile for at least eight days to effect good organic breakdown. A maturation period of two months is recommended.

Other key factors for successful composting are that the carbon:nitrogen ratio should be about 30:1, the moisture content 50%−60% and pH 6−9. Woodchips have a C:N ratio of about 150:1 and compost very slowly. Farm slurries and sewage sludge by contrast have a C:N ratio of about 10:1 and tend to release the nitrogen as ammonia; this leads to odour problems. A bulking agent is therefore a necessary addition, and woodchips, straw, shredded paper and the biodegradable fraction of domestic refuse have all been used, good mixing being essential. Ultimately, the compost is screened to recover any unused bulking agent: typically, woodchips can be reycled three or four times.

Materials for composting must have a moisture content of about 60%; sewage sludge will require dewatering to a dry cake first, and/or be intimately mixed with a dry bulking agent. Lime-treated materials with a pH over 9 will compost slowly and can ultimately have very high calcium levels, which inhibits plant growth. In a compost pile, the pH starts at about 6.5, falls due to acid prôduction and eventually rises to about 8.

Composting can either be conducted in windrows, which are covered and mechanically turned in large facilities, or in aerated piles formed over perforated pipework from which air is drawn by fan. At least one turn is necessary to avoid temperatures reaching 70−80°C, enough to kill most micro-organisms including the beneficial thermophilic bacteria & fungi. Some run-off will occur, which is likely to be highly polluting and may contain metals originally in the material.

If plenty of space is available and possible odour generation will not cause a local nuisance, this is a very 'green' method of sludge treatment. The product, when further mixed with straw or selected household solids, has potential as a saleable product in DIY and garden centres. Several of the UK Water PLCs are currently conducting trials.

Although composting has a considerable history, reliable performance and the factors that will ensure it are not well documented, although Carroll *et al.* (1993) give a useful overview. A small pilot plant is therefore recommended before any large-scale operations. Compost production costs in the UK are reported to be £10−20/t DM.

5.5.3 Mechanical dewatering

A number of methods are available, as outlined below. In many cases, some degree of simple consolidation or thickening to about 4% DM is a preliminary requirement if cake of 30% or more is required (see Section 5.5.1). All methods involve mechanical handling and pumping, and the application of pressure, centrifugal force, shear or heat. A cake of between 25%–50%, depending on feed sludge composition, may be expected. Most equipment manufacturers are very willing to carry out treatability and dewatering evaluation of sludges, or conduct on-site trials, for which purposes trailer-mounted mobile test rigs are often available.

Almost every type of sludge, except those with a very high fibre content, will require some initial chemical conditioning and possibly pH correction before mechanical dewatering, to assist the separation of water from solid particles by reducing chemical and physical attractive forces. Lime and ferrous sulphate (called copperas in the trade) were widely used in sewage treatment until recently, and, being cheap or scrap, by-products from some processes still find application in industrial presses. Both are mixed with the sludge in a tank with slow stirring – the lime as a slurry (typically 10%–15%), the copperas as a 20%–30% solution. Dose rate will require individual trials; the dry lime mass required is often 10%–15% of the DM mass of the sludge (copperas 2%–5%). The conditioned sludge is then pumped straight to the press.

Ferric chloride and aluminium sulphate are other inorganic conditioning alternatives, but a wide range of organic polyelectrolytes are now available. Dose rates for polyelectrolytes to thicken sludges are typically 2–5 kg dry polymer/t of dry sludge solids and thus the chemical is often purchased in 50 kg sacks, and requires little storage space. Costs for treatment and ancillary dosing equipment are usually less than for inorganic chemicals.

There are a number of polyelectrolyte make-up systems available; Allied Colloids Ltd offer a typical product comprising a powder screw feeder and air blower with mixing and dilution tanks to make a 1%–2% solution. In-line injection then occurs through a dispensing pump coupled to sludge flow rate, and on the delivery side of the sludge pumps feeding the dewatering press equipment. This is a popular configuration, simple, economic of chemical and accurate. A 30%–35% cake is readily obtained by most methods, and those with a high fibre/lime/ferrous sulphate content will form a dense, cardboard-like product that usually does not absorb water.

Lime has been widely used historically for sewage sludge treatment. The considerable pathogen kill rate (99% for tapeworm cysts) and a reduction in odour due to the release of ammonia from the very dry, alkaline cake produced has made it very acceptable to farmers with acid soils as a long-term fertiliser/conditioner.

Many polyelectrolyte-treated sludges of organic origin tend to form jelly-like masses which can smell and subsequently reabsorb rainwater, a factor to bear in mind if long-term on-site storage before disposal is contemplated. Pathogen kill rates are low – less than 10%.

Where incineration is the ultimate fate of the cake, inorganic treatment chemicals will reduce the calorific value of the sludge by 10%–15% and increase ash volumes, whereas polyelectrolyte enhances it by 5%–10%. The differences may significantly affect thermal efficiency and running costs. Only polyelectrolytes are therefore used to dewater sludge in present incinerator designs.

It must be remembered that the filtrate produced by dewatering organic sludges will be highly polluting and odorous, and require some form of biological treatment. At sewage works, the liquor is often picked up with other works drainage and returned to the works inlet. Ammonia levels can exceed 100 mg/l, not a problem if the works is correctly sized. However, where the works is fully loaded organically, strong sludge liquors can overload the aeration stage causing partial loss of nitrification and degeneration of sludge settling properties. If the liquors are produced sporadically or from batch dewatering processes, a balancing tank providing a constant but low flow for treatment can be the solution.

In difficult circumstances, where there is neither diluting flow nor a treatment plant, separate pre-treatment comprising settlement, air stripping or high-rate filtration can be used, and the industrial producer will often find this helps to reduce trade effluent charges where the liquor enters the foul sewer. The main point to be aware of is odour production.

Plate press

This consists of a series of substantial iron plates up to $2 \, m^2$ and each covered by a woven terylene/polypropylene cloth. A central hole in each plate admits sludge under pressure, which then spreads across the plate cloth surface, dewatering to a cake as it does so. The filtrate passes through the cloth and the accumulating solids, which themselves assist the filtration process, collecting in a series of channels draining to holes on the outer edges of the plate. From these, the filtrate discharges from one end of the press, or may pass down through the hollow interior of the plate for discharge at the base through a threaded tap.

The whole press is clamped tight by hydraulic rams or a screw during operation, which restrict the expansion of the press as it fills. Ram pumps are commonly used to inject the sludge at pressures up to 15 bar; smaller presses may use peristaltic pumps.

There has been a trend in recent years to increase operating pressures and also to incorporate rubber membranes between pairs of plates. These are inflated with either water or compressed air after the main press fill has occurred. Extra water is squeezed out and total press time substantially reduced. Both sewage sludges and thin slurries dewater better in membrane plate presses, and the pressing cycle time is often halved, allowing two pressings per working shift.

Figure 5.3 shows a cross-section through a large plate press typical of the water industry and supplied by Edwards & Jones Ltd, while Figure 5.4 shows a close-up of the action of the membrane press. Figure 5.5 shows a small industrial press used to treat sludge from photo-etching work, high in iron and copper and disposed of by skip as a controlled waste.

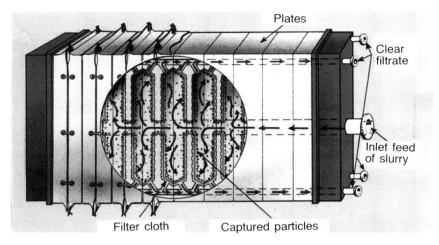

Figure 5.3 Cross-section through a plate press. (Courtesy: Edwards & Jones Ltd.)

In normal operation, the sludge pumps are run until a fixed pressure is reached, and the sludge left to dewater for 3−6 hours. The pressure is then released, the press opened and the cake discharged onto a conveyor or to some form of transport. As it is a batch process, pressing can often be carried out well within the working day and with minimal supervision. If it precedes a continuous process like bagging or incineration, storage hoppers will be needed; 30%−40% DM cake is readily achievable, which is dry enough to pick up in sheets and dispose of by skip.

The cloths will, in time, become blinded by 'fines' and calcium scale in hard water areas. Rag, grit and debris can also be a problem, and pre-screening of the feed sludge may be found necessary. High pressure cleaning is normally effective, about monthly in most cases. In extreme cases the cloths will need to be removed from the plates and soaked in acid; the use of lime as a conditioner encourages this problem. The cloths should last over two years, and represent the most expensive maintenance item. They must be regularly checked for splits, as the pressure will induce a weak plate cloth to 'blow', which is neither safe nor pleasant! The amount of mechanical handling and scraping to remove sticky cake will dictate cloth life, and only blunt tools must be used.

As plate presses perform better with a high fibre sludge, consideration should be given in industrial operations to incorporating any available paper fibre with the sludge, including the reduction of scrap paper and cardboard from the offices to a fibrous slurry. This will solve two disposal problems, and cut down on treatment chemical dosing rates. Extreme cases may produce a cake over 50% DM, and with a texture like insulation board, simplifying handling and reducing disposal volumes still further.

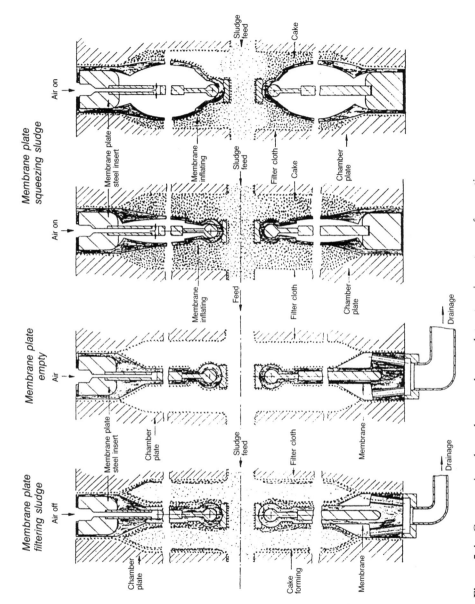

Figure 5.4 Cross-section through a membrane plate at various stages of operation. (Courtesy: CIWEM.)

Figure 5.5 Small industrial plate press.

Plate presses are one of the cheapest dewatering techniques to operate in terms of chemical conditioning. For a plant using polyelectrolyte dosed at 1.5 kg/t DM to produce a 30%–35% DM cake, costs are about £6/t DM. The manual labour requirement during cake discharge will offset this economy against the continuous processes described next.

Belt presses

Belt presses offer continuous cake production with minimal labour, and are very popular. That produced by Simon Harley Ltd is one of a number in the UK market. Sludge is exclusively conditioned by polyelectrolyte in this method, a dilute solution being injected on the delivery side of the feed pumps discharging to a loading hopper or gravity chute. The aim here is to avoid shearing the delicate flocs formed, and load the continuous belt press as gently as possible. Polymer make-up, outlined in Section 5.5.3, is automatic, the equipment being sized to supply one day's worth of belt operation.

Figure 5.6 shows a cross-section through a belt press. Water is squeezed out of the sludge by the pressure applied by a series of rollers, the belt describing a circuitous path over and between a number of them. In some designs, shear is applied by rotating the rollers at different speeds and inducing sideways movement. These actions are provided hydraulically by an on-board pump, making pressure variation possible, although once commissioned by the supplier for the best results, most operators leave the initial settings alone.

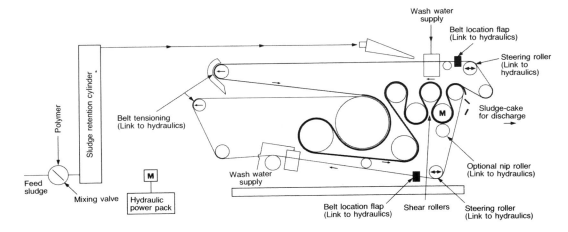

M = motor

Figure 5.6 Cross-section through a filter belt press. (Courtesy: CIWEM.)

High pressure water washing of the belt takes place as it returns to the sludge pick-up trough. Continuous production of cake will normally require a conveyor to transport the material to tip or incinerator. A paddle blender and screw auger compactor can also be used to convey the cake down pipework and reduce its volume, the lubricating properties of the polyelectrolyte conditioner being useful here.

Labour costs are less than for a plate press as most installations are entirely automatic, and will run safely unsupervised, requiring only visual inspection. Chemical costs are higher, reflecting the much shorter treatment time and higher throughput; 3–9 kg polymer/t DM is a likely range to produce 30% DM cake. These dose rates equate to £8–£20/t DM produced. The higher dose will apply particularly to activated/aerated organic sludges, which are always difficult to dewater and where possible should be mixed with raw sludges first. Surplus oxidation ditch sludges often fail to dewater above 22% on a belt press, the end product being gelatinous but not difficult to handle or spread on land.

The screw press

This simple system available from Biwater Europe Ltd, is illustrated in Figure 5.7. Polymer conditioned sludge is pumped to the base of an inclined screw auger and the material is conveyed upwards into a dewatering basket where most of the free liquid drains out. Solids continue upwards to the discharge point, itself restricted to a small orifice to generate back pressure; 25% DM cake is claimed to be achieved with a fairly coarse, raw sludge for which it is suitable, and the operation is enhanced by fibres, hair and even coarse screenings. Surplus activated sludge will produce a thick paste greater than

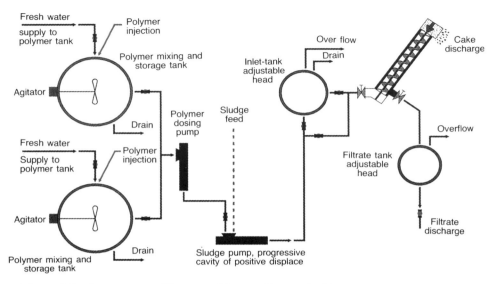

Figure 5.7 Screw press. (Courtesy: Biwater Europe Ltd.)

10% DM. $5\,m^3/h$ of feed sludge can be handled. No actual operating costs are available but £5–10/t is estimated.

Centrifuges

The solid bowl decanter centrifuge has been widely used in industry for some decades, but it is relatively uncommon in wastewater sludge treatment. The advent of cheap polymer treatment chemicals makes it a viable option, producing a 20%–30% DM cake with most materials; 45% DM has been claimed with fibrous sludges. The process is continuous, and has been widely examined as a precursor to incineration. The device is shown in Figure 5.8, and comprises a conical–cylindrical rotor with a screw conveyor inside it.

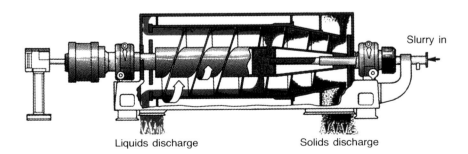

Figure 5.8 Cross-section through a centrifuge. (Courtesy: Alfa Laval Separation Ltd.)

Both rotate at slightly different high speeds in the same direction. Solids introduced are flung onto the rotor periphery, the water forming a pool in the centre and draining over weirs at the wide end of the rotor. The solids are screw-conveyed towards the narrow end of the rotor bowl. Both conveyor speed and the depth of the water pool can be adjusted to maximise performance. Alfa Laval Separation Ltd is one of several suppliers.

A principal advantage is that the equipment is totally enclosed unlike presses. This reduces odour problems and the need to ventilate entire buildings and enabling siting of the centrifuge to be near the sludge source inside the factory. The process is particularly attractive for organic, food-based materials and fine, inorganic sludges. Labour requirements are minimal. Chemical dosing may not be necessary and so costings are not possible to estimate with certainty. Where chemicals are needed, the polyelectrolyte dose is typically 4–8 kg/t DM at 1995 prices.

Sludge dryers

The banning of UK sewage sludge disposal to sea by December 1998 has refocused attention on methods for not only reducing sludge volumes but also converting the material into an easily handled, odourless material that might be sold to farmers, the public through garden centres and even, after sterilisation, as an animal feed supplement.

In the UK, one Water PLC has invested heavily in thermal drying to convert digested sludge to an attractive pelletised product. The process has been used widely in the food and agricultural industries for many years. Much of the development has been in Europe, either as a preliminary to incineration or a 'stand alone' process.

The sludge is dehydrated in a closed vessel by some heating medium and to a temperature that prevents any significant decomposition of the main constituents, reduces water content and bulk by 10- to 30-fold and evaporates the odorous volatile components. The product is pleasant and easy to handle, of calorific value similar to wood (15 MJ/kg) and 90% DM.

Two distinct processes are available:

(1) Direct drying brings the sludge into intimate contact with either superheated steam or hot air which may also convey the sludge. The efficiency of heat transfer is high; the Swiss Combi plant, manufactured by W. Kunz AG, is an example of this (Figure 5.9).
(2) Indirect drying separates sludge and heat source, the former being moved across the drying surface by agitation or scraping. Heat transfer is less efficient but lower air volumes are used that require treatment to remove odours. Seghers Engineering is a supplier of an indirect drier.

In the direct drying method a stainless steel drum is slowly rotated in the horizontal axis, being substantially insulated outside and with vertical diaphragms incorporating central holes and longitudinal plates running between

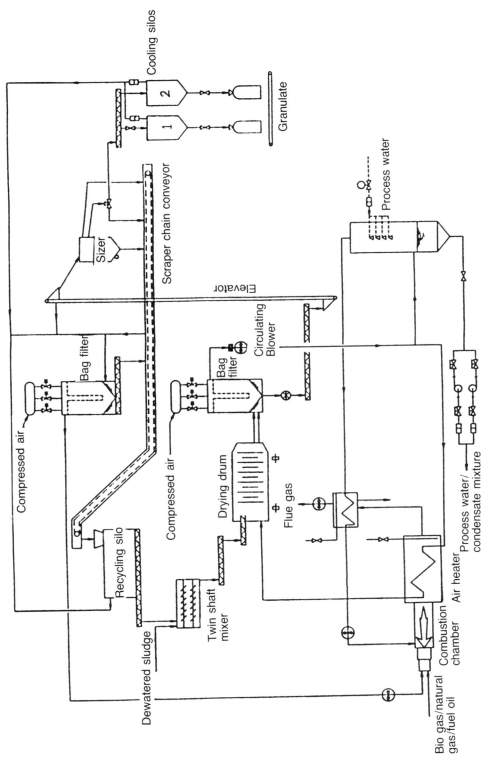

Figure 5.9 Layout of the Swiss Combi sludge drying plant. (Courtesy: CIWEM.)

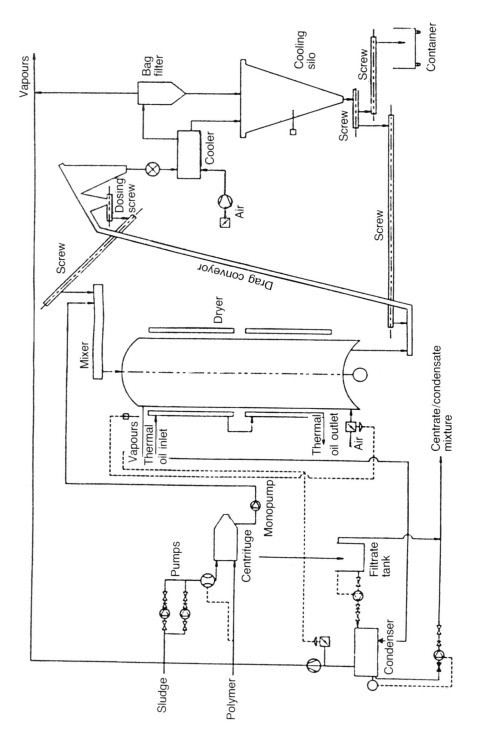

Figure 5.10 Layout of the Seghers sludge drying plant. (Courtesy: CIWEM.)

the diaphragms, inside. Sludge is lifted by the rotating action and the plates, the heated gases driving the granulating sludge forward. Lighter particles pass through quicker but emerge with the same moisture content as the heavier particles, giving a higher operating efficiency than indirect dryers with a fixed residence time for all material.

Heat is supplied by oil, gas or waste heat from an incinerator, where this forms the ultimate disposal route. A boiler heats the recirculating steam/air mixture, the air being drawn from areas of the plant likely to produce odours. This serves the double benefit of using pre-warmed air and providing a post-combustor to remove odours. Air enters the dryer at 450°C.

Sludge entering the plant is dewatered to at least 25% DM, usually in a centrifuge and then mixed with recycled dried granules. After passage through the dryer, a bag filter collects the dry material, now about 90% DM, a screen sizer sorts the granules and returns small and large to be recycled, while granules of 2–4 mm are bagged.

Air/steam from the filter passes through a spray condenser which can be supplied with second quality water or final effluent. The condensate at 90–95°C can be used for heating (digestors for example), the scrubbed air returning to the boiler.

The indirect drying method, as supplied by Seghers Engineering, is shown in Figure 5.10. The dryer drum rotates on a vertical axis and internally contains a number of jacketed, horizontal steel trays. The heating medium circulates at 250°C. Four raking arms move the sludge radially from one tray to the next as it passes from the top feed hopper to the base of the dryer. Similar to the Combi plant, dewatered sludge and recycled granules comprise the feed material; process air and evaporated water pass through a condenser, the liquor emerging at 70°C. Odours would be extracted from the plant and passed through an after-burner, but in the installation illustrated, the dried sludge is incinerated and thus no separate boiler is required to heat the thermal oil passing through the tray jackets, foul air being fed direct into the incinerator.

Operating experience has shown that the Combi plant consumes about three times more electricity than the Seghers plant but is more thermally efficient, and thus overall costs are similar. As the end product has marketable value as a fuel substitute for coal or wood, or as a fertiliser, the operating costs of such plants will depend upon the sludge disposal route, the type of fuel used to fire the boiler, the presence of any 'green premium' incentives to use one type of fuel, the potential use of waste heat and the feed material composition.

Present stand-alone installations at sewage works have used the Combi process to generate a saleable product, the Seghers Engineering plant being tied to incineration. Gross (1993) gives an interesting account and comparison of the two methods including wider environmental issues such as noise, dust, traffic, atmospheric emissions and visual impact.

5.6 Digestion

This is the slow degeneration of the organic content of sludge by obligate anaerobic bacteria to simpler compounds – ultimately carbon dioxide, water and anions (nitrate, sulphate and phosphate) although the process never achieves this total breakdown. At ambient temperatures below 10°C, the reaction rate is very slow, and a lagoon of undisturbed sludge might take many years to undergo significant aerobic/anaerobic digestion breakdown. In warm, temperate and tropical climates, 6 months may produce a well-matured, earthy product. Mesophilic anaerobic digestion is conducted in a closed tank and ideally at 37°C. The process is fragile and relies on three groups of organisms.

Group 1 hydrolyses organic polymers, lipids, monosaccharides and amino acids.

Group 2 ferments the breakdown products to simple organic acids, e.g. acetic, propionic, and comprises both facultative and obligate anaerobes – the acidogens, e.g. *Clostridium*, *Delsulphovibrio*, *Actinomyces*, *Staphylococcus* and *E. Coli*.

Group 3 – the methanogens – converts the H_2 and acetic acid formed to CH_4 and CO_2; *Methanobacterium*, *Methanobacillus*, and *Methanococcus* are commonly present. Group 3 can only use a few substrates, including CO_2, H_2 formate, acetate, methanol, methylamine and CO and has a low growth rate.

Groups 1 and 2 must be in dynamic equilibrium while Group 3 will not tolerate DO, toxic metals or sulphides. The pH must be in the range 6.6–7.6 and never less than 6.2.

The digestor gas produced is typically 65%–70% methane, 30%–34% carbon dioxide, with 200–500 mg/l hydrogen sulphide and traces of other sulphur compounds. This collects either in the space above the digesting sludge, the tank having a rise and fall roof, or in a separate holder.

Methane gas forms explosive mixtures with air in the 5%–13% range and all equipment must be spark free. A number of safety regulations will apply regarding the storage of digestor gas, and each reader must address these as they apply to his or her country or state. Digestor gas represents a considerable energy source, equivalent to bottled or municipal piped gas supply. The collected gas is burnt in a boiler to keep the digestor warm and the excess put to further heating or power generation purposes.

Digestion is one of the few sludge treatment processes in which a significant reduction of pathogens is possible. The ova of the beef tapeworm *Taenia saginata*, the nematode (roundworm) *Ascaris suum* and *Salmonellae* are industry 'standards' and anaerobic digestion reduces all of these by 90%+. This makes it an attractive process for the farmer intensively rearing livestock and with a lot of liquid slurry to treat before spraying back on fields; 10- to 100-fold reductions of numbers of Salmonella, Enteroviruses and *E. Coli* are usual.

Organic sludges containing nitrogen and phosphorus will digest to an earthy consistency and are an ideal slow-release fertiliser. Digested sludge does not dewater as well as raw material but a 25%–30% cake should be achieved by a belt or plate press. As the volatile content will have dropped by about 40% release trapped gas and avoid dead pockets. Mechanical stirrers, circulating pumps and gas compressors are used, and can be run continuously, but more often for 4–8 hours/day after a new batch of sludge has been added.

The process is operated on a fill and draw basis, residence times being 15–40 days, although some intensive systems can operate at 10 days. Normally the digestor is fed with fresh sludge daily, although not feeding over the weekend or short holiday does not usually cause problems. About 5% of the digestor contents are replaced at each feed. Daily digestor loadings of 0.5–1.6 kg volatile material added/m^3 digestor capacity/day for standard rate and up to 4.8 kg for high rate are typical. For standard rate digestion, 50–80 l/head/day digestion capacity is allowed. The feed sludge should be 2%–4% DM; sludges over 7% are difficult to pump and may produce too much gas for safe handling. The organic biodegradable content of the sludge should exceed 70%.

Simple analysis for DM and O & V% content, and for the presence of any inhibitory substances that circumstances may indicate are present in the sludge, should be carried out. If this facility is not available a small laboratory scale digestor may be run in parallel with the main unit, and may be fed in advance of the main unit with each sludge feed batch. If there is a sudden reduction in gas production, investigate the cause and do not feed the main digestor!

Commissioning a digestor entails the purging of the whole tank and any gas holder with nitrogen. The sludge should be thinned to 1%–2% DM, seeded if necessary with digested sludge, pumped in and heated to 30°–40°C. An external heat source will be required at first.

The inexperienced are strongly advised to employ a specialist firm, the original suppliers or experts in digestion to commission the process. Industrial sludges often need seeding with digested sewage sludge to add sufficient bacteria, and sewage sludge will need thinning with effluent or sewage initially. Tankering costs must be anticipated.

As the process matures – it may take several months – the digestor contents are slowly mixed. Methane-producing bacteria will preferentially breed, utilise the sludge solids as a food source and produce considerable volumes of methane and carbon dioxide. Once the process is established, fresh sludge is admitted on a daily basis (though never more than about 5% of total digestor volume), and a corresponding amount of digested sludge withdrawn. The pumps can be used to aid mixing, or a mechanical stirrer or gas mixer can be used.

Although the normal sludge retention time is 15–25 days, there have been experiments with much shorter time periods in an attempt to achieve greater throughput, but these experiments need rigorous control and are only for the digestion will virtually cease. In cold climates, it can be beneficial partly to bury digestors, and many modern prefabricated designs are well lagged but thin-walled.

Mixing is essential to maintain the process by ensuring even heat distribution, release trapped gas and avoid dead pockets. Mechanical stirrers, circulating pumps and gas compressors are used, and can be run continuously, but more often for 4–8 hours/day after a new batch of sludge has been added.

The process is operated on a fill and draw basis, residence times being 15–40 days, although some intensive systems can operate at 10 days. Normally the digestor is fed with fresh sludge daily, although not feeding over the weekend or short holiday does not usually cause problems. About 5% of the digestor contents are replaced at each feed. Daily digestor loadings of 0.5–1.6 kg volatile material added/m^3 digestor capacity/day for standard rate and up to 4.8 kg for high rate are typical. For standard rate digestion, 50–80 l/head/day digestion capacity is allowed. The feed sludge should be 2%–4% DM; sludges over 7% are difficult to pump and may produce too much gas for safe handling. The organic biodegradable content of the sludge should exceed 70%.

Simple analysis for DM and O & V% content, and for the presence of any inhibitory substances that circumstances may indicate are present in the sludge, should be carried out. If this facility is not available a small laboratory scale digestor may be run in parallel with the main unit, and may be fed in advance of the main unit with each sludge feed batch. If there is a sudden reduction in gas production, investigate the cause and do not feed the main digestor!

Commissioning a digestor entails the purging of the whole tank and any gas holder with nitrogen. The sludge should be thinned to 1%–2% DM, seeded if necessary with digested sludge, pumped in and heated to 30°–40°C. An external heat source will be required at first.

The inexperienced are strongly advised to employ a specialist firm, the original suppliers or experts in digestion to commission the process. Industrial sludges often need seeding with digested sewage sludge to add sufficient bacteria, and sewage sludge will need thinning with effluent or sewage initially. Tankering costs must be anticipated.

As the process matures – it may take several months – the digestor contents are slowly mixed. Methane-producing bacteria will preferentially breed, utilise the sludge solids as a food source and produce considerable volumes of methane and carbon dioxide. Once the process is established, fresh sludge is admitted on a daily basis (though never more than about 5% of total digestor volume), and a corresponding amount of digested sludge withdrawn. The pumps can be used to aid mixing, or a mechanical stirrer or gas mixer can be used.

Although the normal sludge retention time is 15–25 days, there have been experiments with much shorter time periods in an attempt to achieve greater throughput, but these experiments need rigorous control and are only for the

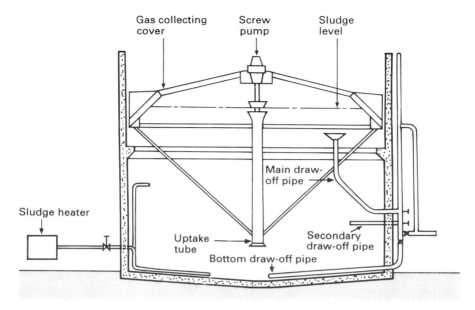

Figure 5.11 Cross-section of an anaerobic sludge digestor.

experienced operator (Noone & Brade, 1982). A cross-section of a digestor and ancillary equipment is shown in Figure 5.11.

A healthy digestor will produce gas continuously with little volume variation/day, but the literature reports a wide production range. For sewage sludge, gas yield is about $0.25\,m^3/kg$ DM sludge added. Other ways of expressing gas yield are in relation to the volatile matter or O & V content of the sludge; $0.35\,m^3/kg$ O & V added and $0.75\,m^3/kg$ O & V removed are typical. Quoted gas production values, measured in l/capita/day (or for farm digestors an animal equivalent), vary widely between 2 and 35.

Calorific value of the gas is typically $25\,kJ/l$, while natural gas is about $35\,kJ/l$. It is used primarily to fire a boiler to maintain the digestor temperature by circulating hot water through the tank. A few designs withdraw sludge, heat it in an exchanger and return it.

For many years, larger sewage works have relied on digestion to provide fuel to generate electricity from gas engines, heat the site buildings and (in a few examples) compress the surplus and run vehicles or sell it on to industry. The UK alone has a number of sites where the incoming sewage flow exceeds $50,000\,m^3/day$ and sludge digestion has been maintained for 60 years, the works being net exporters of energy.

Recently, smaller combined heat and power (CHP) units have made this a feasible option on sites of $2000-3000\,m^3/day$ flow. In common with larger installations, the engines will run on digestor gas or diesel fuel; engine heat warms the sludge through an exchanger, and the electricity is sold to the grid under the 'green tariff' arrangement whereby the sell price exceeds the bought-in price by 2−3 pence/kWh.

A common unit size is 0.5 MVA, and a contributing population of 100 000 might be expected to generate £150 000/annum at 1994 electricity prices.

The hydrogen sulphide content of digestor gas varies between 100 and 300 mg/l and can be much higher. As this is a source of corrosion in gas engines, forming sulphuric acid, the economics of CHP where the digestor gas contains more than 2000 mg/l H_2S are questionable. Fulton (1991) gives an account of how this problem was overcome by adding ferric chloride to the raw sludge before digestion, and also suggests that ferric chloride might form the heart of an effective H_2S gas scrubbing system.

In summary, digestion is a well-practised art in wastewater treatment worldwide. Not only does it realise the potential energy of sludge and allow self-sufficient operation, but it also converts a foul-smelling material with restricted disposal options into an earthy, watery fertiliser, while reducing the volume for disposal by about 30%. There is thus considerable documentation of the process, and the reader is referred to the professional journals; the Chartered Institution of Water and Environmental Management (CIWEM) is a particularly good source.

The fragility of the process dictates good control of feed sludge quality and the water industry operates trade effluent control and tight limits on a number of industrial chemicals. Table 5.2 lists some of the common substances and their concentration in sludge which will inhibit anaerobic digestion.

Reference was made in Chapter 1 to the importance of effective grit, rag and detritus removal. During digestor operation, these materials will block pipes and slowly fill the tank(s) with an inorganic mass, reducing available digestion volume, gas production and retention time. It is not unknown for a 20% reduction in actual tank capacity to be achieved in five years. The resulting shutdown, venting, digging out and recommissioning (including purging with nitrogen) is an expensive task taking all the health and safety requirements into account. Other maintenance areas that must be anticipated include a gradual corrosion of metalwork − mixer tubes in the tank are particularly prone to this − scaling of boiler tubes and heat exchangers and corrosion of gas compressors where used.

Anaerobic industrial sludge digestion is less widely practised, although farm slurries are particularly suited to digestion and a number of small package plants are now available which allow surplus gas to be used to heat buildings. Farm Gas Ltd supplies a small-scale package plant which has also been used on rural sewage works. The food/dairy industries are also good candidates for considering digestion providing operations are on a sufficient scale − 1 m^3 wet sludge/day is a probable practical minimum.

A primary consideration for the industrial user is to ensure feed sludge quality, i.e. a sludge of consistent organic content and devoid of toxic components. Often this is difficult where production runs and batch processes are

Table 5.2 Common substances inhibitory to anaerobic sludge digestion.

(a) Toxic metals

Metal	Concentration range (mg/kg DM)
Zinc	30–140
Nickel	7–50
Cadmium	200–800
Copper	500–3000

(b) Chlorinated hydrocarbons

Solvent	Concentration (mg/kg DM)
Chloroform	15
Trichloroethane	20
1,1,2-Trichlorotrifluoroethane	200
Carbon tetrachloride	200
Trichloroethylene	1800
Tetrachloroethylene (Perchloroethylene)	1800

(c) Anionic detergents
1.5%–2.0% on DM gives 50% inhibition.

Notes to Part (a):
1. Values are for a 20% reduction in gas yield.
2. Effect varies with solubility and therefore pH and sulphide concentration.
3. For mixtures, the effects are additive on an equivalent weight basis or meq (milligram equivalent weight)/kg DM. Thus:

$$K(\text{meq/kg}) = \frac{(Zn)/32.7 + (Ni)/29.4 + (Pb)/103.6 + (Cd)/56.2 + (Cu)/47.4}{\text{Sludge solids concentration in kg/l}}$$

If K is 400 meq+, there is a 50% chance of digestor failure.
If K is 800 meq+, there is a 90% chance of digestor failure.
If K is <160 meq, there is a 90% chance of digestion being unaffected.

Note to Part (b):
1. Values are for a 20% reduction in gas yield.
Values are Courtesy of CIWEM.

erratic without space-consuming sludge storage. On-site skill to safely manage the plant is also essential. The methane gas produced is very flammable and explosive, and commissioning or restarting a plant is definitely a matter for the suppliers or independent experts. Storing methane gas close to buildings or roads is subject to planning restrictions and fire regulations.

In the author's experience, the scale of sludge production and its variable composition are not attractive for digestion in many industrial operations; farms are an exception because here the digested material can be disposed of without any further treatment or significant handling.

The energy extracted will not warrant the capital cost of the plant in most countries where the price of electricity is less than 10 pence/kWh unless the digestor gas has another application, e.g. as a chemical feedstock or an alternative fuel for vehicles during fuel shortages. Sludge volume reductions can be better achieved by mechanical dewatering. The sensitivity of digestion to toxic shock loads which many factories will not have the analytical capability to detect (and about which, in the cocktail unique to each factory, few documented accounts are likely to exist) also favours a cautious approach to digestion in industry where monitoring resources are very limited.

Digestion costs in the water industry are quoted over a wide range. Operating costs/head per annum are between £1 and £4; total costs/head per annum range between £20 and £60, underlying the capital costs associated with digestor construction. The availability of relatively cheap electricity in Europe in the last 20 years and security of supply has therefore eroded some of the power generation advantages of digestion.

5.7 Incineration

Incineration will become more widely used as a technique in the next 20 years, particularly in the UK and other islands denied sea disposal of sludge. Its main advantages lie in the complete destruction of organic matter, the ash being inert and usually less than 25% of the original sludge volume. This can be particularly useful for industrial sludges containing toxic, poorly biodegradable components. The ash concentrates metals, and although rarely engineered into present designs, the opportunity to recover them by acid stripping or ion exchange. At present, such ash is considered inert and only under acidic and anaerobic ground conditions are metals likely to be leached out. Disposal to a controlled tip is permitted at present. If sludge is deemed to be a hazardous waste in future, ash tipping outlets will become much more restricted.

The main area of concern for present operators and new designs is the ever-tightening emission standards reviewed below in Section 5.7.4. All designs have considerable safety interlocking, automatic shutdown and computer control. Skilled operators/supervisors are required to run such a plant with mechanical and electrical back-up to cope with the complexities of maintenance.

With few exceptions, most incinerators are of the fluidised bed variety, as illustrated in Figure 5.12. The one depicted here is designed to burn 7–8 t of 30% DM cake/hour. A flowsheet for a similar installation that can also take screenings and imported sludge cake is shown in Figure 5.13. This is manufactured by Dorr Oliver Company Ltd.

In summary, the process comprises initial drying to produce a cake, followed by incineration at 850–900°C, the success of the latter being dependent on the production of sufficiently dry cake to ensure autothermic operation and so avoid the costly supplementary fuel (usually oil or gas).

Belt presses, plate presses and centrifuges all appear in present designs and the DM obtained by working documented examples varies between 27% and

Sludge Disposal and Treatment 141

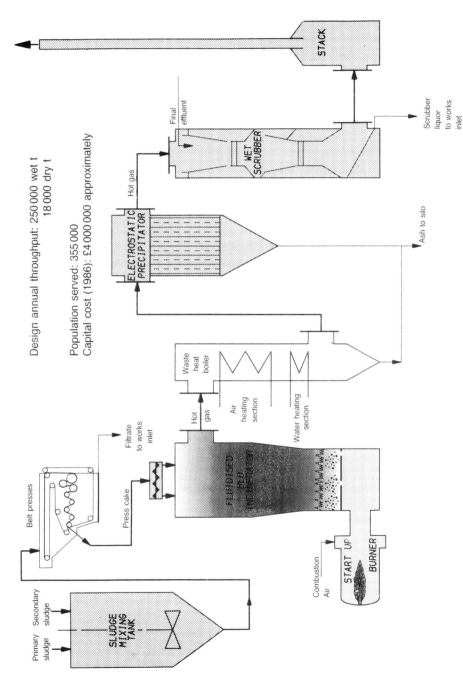

Figure 5.12 Layout of a fluidised bed sludge incinerator.

Reproduced from a paper by J. Hudson, PhD, MIWEN & R.J. Walker, BSc, MICE, MIWEM, published in the CIWEM Yearbook by FSW Publications, Norwich.

142 Sewage and Industrial Effluent Treatment

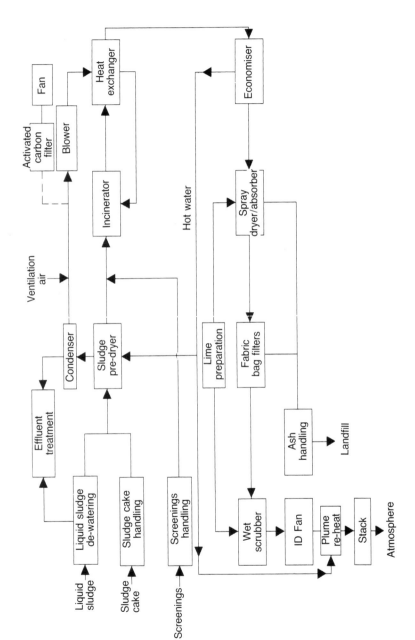

Figure 5.13 Flow sheet for sludge/cake/screening fluidised bed incinerator. (Courtesy: Dorr Oliver Company Ltd.)

35%. The filtrate produced requires biological treatment, and on a sewage works it is returned to the works inlet. Sewage sludges of 75%–80% O & V content, and treated with a polyelectrolyte conditioner, will then have a calorific value of 20+ MJ/kg DM; 25 MJ/kg DM and a feed sludge of 30% DM and 25% ash content (sewage sludge is typically 24% ash content) is a common design value.

DM, O & V% and subsequent calorific value are the three important parameters for ensuring autothermic operation, a fundamental requirement of most contemporary incinerators designed for sewage sludge disposal. A marginal increase of either DM or O & V% will permit exothermic operation, and the two sludge incinerators being built by Thames Water PLC in East London are designed for this state, the excess heat being used to raise steam, generate electricity to power the activated sludge aeration plant and provide local heating.

The variability of sewage sludge composition will mean that at some locations, exothermic operation will be very difficult to achieve, and continuous autothermic operation unlikely. Industrial sludge composition in batch processes can be more carefully controlled by segregation. Although often of lower organic content, the material can frequently be dried to 40%+ DM to achieve autothermic operation.

5.7.1 Operation

As a preliminary, it is essential to provide adequate wet sludge and cake storage to cover periods of incinerator breakdown or maintenance; 28 days has been allowed in recent designs burning 10 t sludge DM/hour. Some initial blending and dewatering is in any case highly desirable so that the dewatering system can be supplied with a consistent sludge of at least 4% DM. In the system illustrated, 7%–8% DM is expected.

The cake falls from the presses onto a chain conveyor and a screw conveyor transports the fragmented material to two loading ports on the top of the incinerator. Earlier designs injected the sludge cake from the side via screw pumps about 1 m above the fluidised sand bed. The conveyor can be reversed so that cake can be disposed of without incineration, a useful facility covering breakdowns, and during commissioning when feed rate adjustment is necessary. The incinerator comprises a steel drum, and for an installation designed to burn 7–8 t of cake (2–2.5 t DM)/hour, stands 7 m high and 5 m in diameter. It is lagged internally with fire bricks.

A fluidised sand bed of about 10 m^3 is maintained at 750–800°C, 2 m above the base by external blowers discharging through an arched crown of air bricks. Fresh combustion air is drawn from the building and sludge tanks so that a slight negative pressure is generated and odours are effectively contained within the incinerator. Preheated air from the heat exchanger is mixed with this to achieve fluidised air at about 650°C.

The sludge burns completely in less than two seconds, although most incinerators are designed with an average retention time of four seconds in the reaction zone. Temperatures towards the top of the incinerator approach 900°C, and a water spray is available for quenching. Temperature is controlled by sludge feed and fluidised air rates; an excess air or oxygen monitor controls the latter and maintains 8% excess O_2. Automatic shutdown occurs below 4%.

Hot gases, excess air and ash enter a heat exchanger and cool to 250–300°C. Air-to-air tubes are used to preheat the fluidising air, and water-to-air tubes provide hot water for local heating and to reheat the stack gases at the chimney and avoid any visual plume of water vapour. The example illustrated in Figure 5.12, extracts 2.8 MW of heat. The most recent designs generate electricity from steam turbines and European versions supply district heating systems with hot water.

An electrostatic precipitator removes 95% of the ash and dust particles which become highly charged and attracted to collecting electrodes. These are periodically vibrated, the ash falling into a hopper and being blown by compressed air to a silo. About 1 t of ash is collected for every 4–5 t of DM burnt, and the material taken to a controlled tip. Hot gases enter a wet scrubber. The liquor used can be a good quality effluent and removes the remaining 5% of ash and the gases hydrogen chloride, hydrogen fluoride and sulphur dioxide. A pH of 6.8 is maintained by sodium hydroxide to ensure the gases dissolve. Alternatively, a lime slurry spray is used. Scrubbing is often a two stage operation, the second stage cooling the gases to 50°C. The liquor, which will often have a high COD and need biological treatment, is returned to the works inlet.

The stack gases are fully saturated after scrubbing and if released would generate a visible white plume. As an indication of the change in public perception of pollution, this sight was once thought attractive. Now, quite the reverse is true, and heat recovered from the exchanger is used to warm the gases at the base of the stack to about 130°C and prevent any visible discharge down to −5°C. This is a requirement stipulated by Her Majesty's Inspector of Pollution (HMIP) in the UK. A stack height of 40 m is dictated by the planning authorities in the UK.

5.7.2 Ash composition

Ash from sewage sludge incineration has the consistency of fine silt and various experiments have been conducted to find a use for it. The metal content, being some four to five times that of the sludge, may make extraction an attractive proposition for industrial sludges containing singular elements, but the mixture of components can be problematic to separate from sewage sludge. The metal is tightly bound within the inorganic matrix and Table 5.3. gives a typical analysis. The toxic metal content of this ash is, in the author's experience, unusually low and total mixed metals of 2.5% are likely in heavily industrialised areas.

Table 5.3 Ash composition from a sewage sludge incinerator.

Constituent	% on dry weight of sample	mg/l
SiO_2	54.9	
Al_2O_3	18.4	
P_2O_5	6.91	
Fe_2O_3	5.83	
CaO	5.43	
K_2O	1.86	
MgO	1.27	
TiO_2	1.06	
Na_2O	0.93	
SO_4	0.46	
Cl	0.30	
BaO	0.18	
Cr_2O_5	0.11	
SO_3	0.09	
SeO	0.03	
Trace elements	0.12	
Loss on ignition	1.86	
Total	99.74	
Copper		650
Zinc		450
Nickel		100
Cadmium		11

Reproduced from a paper by J. Hudson, PhD, MIWEM and R.J. Walker, BSc, MICE, MIWEM, published in the CIWEM Yearbook by FSW Publications, Norwich.

Research is currently being carried out at Brunel University, Uxbridge into electrolytic metal extraction methods in soils and ash. In view of the increase in ash volumes produced as incineration becomes more popular, the limited number of landfill sites and the potential for some leaching of metals to take place in unfavourable acidic and anaerobic conditions, it is likely that some incinerators will see metal recovery fitted retrospectively. Alternatively, a severe tightening of trade effluent limits for metals may be imposed, as the collection of volatile metals (e.g. Pb, Cd, Hg) that can be conveyed right through the incinerator and are controlled by the EC Directive governing municipal solid wastes (89/369/EEC) (1989) and levels in the stack gases is expensive – £2 million for an incinerator burning 24 000 t DM/year.

5.7.3 Maintenance and costs

Fluidised bed incineration is a robust, quick and reliable process that can be started and stopped in 10–15 minutes. The author's experiences with the plant at Esher (Dickens *et al.*, 1980) indicated that the robustness and reliability of the control system and sensors critically dictated overall performance and throughput. On the mechanical side, achieving consistent cake quality was difficult at times. After six years, the fire brick lining was wearing thin and generating hot spots, and a costly partial replacement was envisaged.

Present accounts of incinerator operation describe minor problems associated with sludge dewatering, grit and screenings causing conveyor wear, corrosion and excessive expansion at joints around the heat exchanger and ash pumping; 80%–90% plant availability is a common level, but includes initial teething problems. Information on operating costs is sparse; £50–60/t DM burnt for an autothermic plant appears average. Capital costs are high: most incinerators designed to burn 10+ t/hour cost £20–£30 million including the cost of an environmental assessment and compliance with the most stringent emission standards.

Much smaller packaged industrial incinerators are available for less than £500 000. Clearly, if the incinerator runs exothermically and heat energy is extracted, there will be at least one attractive pay-off. The lack of any other form of sludge disposal facility will also dictate incineration as the only viable option so that cost comparisons become irrelevant. The value of metals extracted from the ash must also be considered.

5.7.4 Incineration law and gaseous emissions

These laws are being progressively tightened and a uniform standard has yet to be decided within the EU. The incineration of sewage sludge, in common with other incineration processes, is a prescribed process in Part 1 of the EPA 1990, requiring authorisation from the relevant local authority. Where more than 1 t/hour is burnt, HMIP provides the authorisation in the UK. HMIP issued a guidance note (IPR5/11) (1992) setting out combustion conditions and gas emissions. These are similar to the 1986 German TA Luft standards listed in Table 5.4. along with typical gas stack emissions produced by the Esholt sewage sludge incinerator operated by Yorkshire Water PLC.

EC Directive 89/369/EEC (1989), mentioned in Section 5.7.2, is also likely to be incorporated into English Law. This Directive dictates emission standards for municipal solid waste incinerators almost identical to the 1986 TA Luft standards and it is likely that all UK sludge incinerators will be subject to these emission standards too – new plant immediately and present equipment by 1996. The Directive also dictates monitoring frequencies for a wide range of parameters based on plant capacity. Those of 3 t/hour or more will require continuous monitoring for dust, CO, HCl and oxygen.

An even tougher standard – the German Standard 17 BIMSchV90 – was proposed in 1990. This is listed in Table 5.5 along with a number of emission limits from other European countries and the USA. Of note in the BIMSchV Standard is a dioxin limit of $0.1 \, ng/m^3$. At the operating temperature of 850°C and with 6% excess air in the freeboard at the top of the incinerator, recent reseach has shown that dioxin and furan production is not a significant risk.

Most contemporary incinerator designs are built to achieve this standard,

Table 5.4 Incinerator gaseous emissions and TA Luft standard.

Pollutant	TA/Luft (maximum mg/m^3)	Warren Springs tests* Average test results (mg/m^3)
Dust	30	8.7
Cadmium	0.2	7×10^{-4}
Mercury		2×10^{-2}
Arsenic		2×10^{-3}
Cobalt	1.0	2×10^{-4}
Nickel		3×10^{-4}
Selenium		1.5×10^{-3}
Antimony		9×10^{-4}
Lead	5.0	9×10^{-4}
Copper		6×10^{-4}
Chromium	5.0	3×10^{-4}
Manganese		2×10^{-4}
Carbon monoxide	100	less than 1
Organics (as carbon (C))	20	17.2
Sulphur oxides	100	14
Chlorides (as hydrochloric acid (HCl))	50	9.5
Fluorides (as hydrogen fluoride (HF))	2	0.04

Conditions: dry gas at standard temperature and pressure (STP) and 11% oxygen.
* Esholt incinerator.

although at the time of publication, it is uncertain whether the EU will adopt this ultimately.

A requirement of 85/337/EEC (1985), embodied in UK planning law, is that there be an environmental impact assessment for certain large developments including incinerators. Each site is considered individually, but the Water PLCs have thought it good public relations to carry one out anyway.

5.8 Conclusions

Much of the emphasis in this chapter has been placed on reducing sludge production and disposal volumes, and this approach stems from the certain contraction of disposal routes for all sludges in the future. Landfill sites in Europe are contracting in available volume and number; the operators are charging more, and particularly where the material is deemed hazardous and requires either pretreatment, co-disposal or special site engineering to prevent leaching and groundwater contamination.

The UK government has implemented a landfill levy of £10/t to make other disposal options, in particular incineration, more attractive. As the number of sites diminishes, transport costs will rise out of proportion to others. Poorer

Brunel University, Department of Chemistry, Uxbridge UB8 3PH.
Dorr Oliver Company Ltd, NLA Tower, 12/16 Addiscombe Road, Croydon CR9 2DS.
Edwards & Jones Ltd, Whittle Road, Meir, Stoke on Trent ST3 7QD.
ELE International Ltd, Eastman Way, Hemel Hempstead, Hertfordshire HP2 7HB.
Farm Gas Ltd, Bishops Castle, Shropshire SY9 5AQ.
W. Kunz AG, Dintikon, Switzerland (manufacturer of Swiss Combi Dryer).
Milltronics Ltd, Oak House, Bromyard Road, Worcester WR2 5HP.
Sarlin Ltd, Pumps Division, Highcliffe Road, Hamilton Industrial Estate, Leicester LE5 1TY.
Seghers Engineering, Willebroek, Belgium.
Simon Hartley Ltd, Stoke on Trent ST4 7BH.

Chapter 6
Cesspools, Septic Tanks and Small Sewage Treatment Plant

6.1 Introduction

The homes and workplaces of the great majority of the population in the UK and Europe are connected to main drainage, the construction of the sewerage system and treatment works being an essential element of the development of towns and cities. As expansion has occurred, outlying areas and villages are often incorporated into the drainage system, requiring upgrading and re-building of pumping stations and treatment works and where necessary, the separation of combined surface and foul drainage.

Economics and terrain often define the limits of the sewerage system. It is not uncommon in the UK to find single or short rows of houses and farms with septic tanks only a short distance from large towns, or privately owned small treatment works at conference centres, schools, public houses, hotels or rural camp sites. The go-ahead for a new housing development is frequently dictated by the capacity of the local sewers or pumping station, their re-vamp adding significantly to the costs.

Estimates from a number of surveys carried out by government departments in the 1990s indicate that about 4% of the UK population, equivalent to about 800 000 households or 2 million people are not on main drainage. There is considerable variation between regions with 15% in the South West not connected, but less than 2% in the North East. With sewerage installation costs running at £0.2M–£1M/km, it is likely that these percentages will remain fairly static for the foreseeable future and may rise slightly.

For the purposes of this chapter, cesspools are defined as holding tanks where no treatment occurs and the whole contents are tankered away at regular intervals. The frequency of attendance of a cess-waste tanker to empty a cesspool will depend on personal water use rates, avoidance of waste and leaking drains or a cracked structure that admits ground or surface water.

Septic tanks offer partial treatment, the effluent discharging to groundwater through a subsoil soakaway and occasionally undergoing further biological treatment in purpose-built plant. Solids are retained in the tank and undergo slow breakdown but a tankering service to remove excess sludge at intervals is a normal requirement. Permission to discharge from a soakaway will be needed, issued by the Environment Agency (EA) in England and Wales, the Scottish Environmental Protection Agency (SEPA) or local Authority Public Health Department, and Planning Permission and Building Regulation approval is also usually required.

The volume of septic tanks rarely exceeds $50\,m^3$ (about 240 population equivalent (pe)) and the majority are less than $10\,m^3$. While they can operate for long periods without attention, de-sludging will be necessary at intervals or the effluent discharge quality may suffer and possibly cause local pollution. The Environment Agency can issue an Abatement Order where cesspools or septic tanks cause local pollution to ground or surface water.

Small sewage works offer some form of biological treatment and many new installations are package plant. Relying on a secure power supply, primary settlement, biological oxidation and secondary settlement all occur within one unit.

Alternatively, they may be miniature versions of conventional plant using processes already described in Chapters 1,2 and 4. The term 'small' is used somewhat arbitarily but usually refers to a works serving less than 1000 population equivalent (pe).

Recent years have seen major development of the package small treatment plant specifically designed for underground installation. These range in size from those suited to a couple of houses, pub or small training centre to 200+ population equivalent units. There is also an increasing tendency for the Water Companies to construct underground STWs in sensitive areas, particularly coastal margins, of much larger capacity and often featuring energy-intensive full treatment, forced air ventilation and UV disinfection of the effluent. These are of a sophistication requiring expert knowledge to operate.

The quality of treatment and fabrication of products on the present market for the private owner/operator is very variable and many systems have a constant energy requirement. The advice is to deal with a reputable firm and talk to owners of similar plant to determine reliability, life expectancy, true running costs, maintenance demand and what effluent quality can be consistently achieved.

Compact and/or underground sewage treatment does not mean 'fit and forget' and periodic inspection and maintenance combined with occasional de-sludging are essential. The suppliers are often anxious to secure the maintenance contract but it often pays to find out what level and frequency of service is on offer and an assessment of this and the quality of the proposed installation by an independent expert often repays a day's fee.

A Consent to Discharge or a Descriptive Consent will be issued by the Environment Agency for all small sewage works and samples may be taken at random to check effluent compliance with Consent limits. Again, Planning Permission and Building Regulation approval is also required.

All three systems will be adversely affected by rapid changes in flow rate, often unavoidable but minimised by good design. The discharge of toxic materials including oil, metal solutions, strong acids and alkalis and solvents, excessive over-use of disinfectants, and lengthy power failure at some types of package treatment works will all jeopardise the microbiology and thus treatment efficiency.

The annual operating costs of a cesspool, septic tank or private treatment plant may not be more than the equivalent annual sewerage charge per capita for a mains drainage connection. However the capital costs, compared with laying a connection to the nearest public sewer, may be very different and each new installation needs individual consideration, along with other factors as outlined in the next section.

Finally, it is re-emphasised that cesspools, septic tanks and small sewage works are not maintenance-free and the 'flush and forget' advantage offered by connection to main drainage must now be tempered with knowledge of the consequences of inappropriate actions, maintenance requirements and operating costs. Whatever type of plant is operated, a lack of maintenance will heighten the risk of major failure, flooding and a decline in effluent quality. The consequences of neglect will prove expensive and place legal liability for pollution on the owners. The experience and performance of any maintenance contractors must therefore be regularly checked.

6.2 Choosing a treatment system

For new installations in rural or remote areas without main drainage, it is appropriate to consider first the cost of connection to the public sewer and whether this is even practical or desirable. If the site is 3–5 km away across hilly country where a pumping station will also be needed, the answer is probably no. The 6 houses built on the edge of a village and requiring only 300 m of gravity-laid sewer to connect up to the main sewer provides the converse situation. This issue may also arise if the present treatment facility has failed or proved unreliable and expensive to maintain.

If the circumstances appear borderline, careful long term costing is required. The sewerage connection may be a very expensive one-off but once installed should have no or minimal maintenance costs for many decades. The need for an intermediate pumping station often tips the balance against this option.

As a general rule, cesspools and septic tanks tend to be installed for single houses or small groups of dwellings where space is limited and the maintenance of a sewage works would be financially and logistically onerous. Nevertheless, very small package sewage treatment units to treat sewage from 1–2 houses are now available.

Sewage treatment works are also more suited to larger institutions where plenty of space allows remote siting and on-site maintenance staff have the skills, tools and inclination to effect the right level of attention.

A private treatment system involves capital costs of excavation, laying out and purchase of the tank or plant units. The operating costs will then comprise tankering charges at intervals depending on capacity and use. Some types of small sewage works will need regular mechanical and electrical maintenance and use energy continuously so the power supply must be secure.

New installations of cesspools, septic tanks and package sewage treatment units are subject to Planning Permission and Building Regulations so that effective drainage is achieved. Both septic tank soakaways and small treatment works will require a Consent to Discharge or Descriptive Consent and attract an annual charge levied by the Environment Agency. The effluent from a works will have to conform to Consent Limits and a volume related charge will also be levied where the flow exceeds $5\,m^3$/day (25–30 pe).

The charging formula contains a number of factors and is applied in bands so that there is not a linear relationship between volume and cost. To offset these expenses, no annual sewerage charges will be payable by the owners of a private system. There may be other considerations when choosing a treatment system.

There is no obligation on Water Companies to provide sewerage and connections in the UK if there are practical alternatives, i.e. a septic tank or cesspool. However, if ground water levels are high, installing a septic tank and the soakaway to discharge the effluent may be impossible and a cesspool will have to be substituted.

The close proximity of a water abstraction borehole may also veto the choice of a septic tank and will depend on the current local groundwater policy and soil permeability. In such situations, funding is often available for sewerage projects, but the cost of drainage to larger private housing developments is always paid for by the developer.

To encourage the supplementing of groundwater or stream flows with treated effluent from small works in areas where there may be wide natural seasonal variation, Consent Conditions are locally negotiated and the abandonment of an established works and discharge may be actively opposed by the Environment Agency.

Assuming that a private stand-alone facility is the only option, the main questions to address are:

6.2.1 What flows and organic loads will be generated?

It is usual to allow 180 litres/head/day for private houses and 200 litres/head/day for conference centres and schools with central kitchens. Isolated restaurants, offices and recreational sites with toilets generate 40–100 litres/head/day.

Septic tanks are sized according to the formula;

$$C = (180P + 2000) \text{ litres}$$

where C is the capacity of the tank and P the number of users. This formula derives from historical Codes of Practice but remains appropriate. BS 8301 provides more recent confirmation of the volumes and loadings quoted above which may be used in design and sizing calculations and are equally relevant to small sewage treatment plant. The most important parameters are the septic tank volume and hence the retention time and the volume available for sludge storage and partial degradation.

Large seasonal variations in flow may have to be accommodated by the drainage system at camp sites, hotels, conference centres and large houses with swimming pools which are emptied seasonally. Inlet baffles to septic tanks or treatment works need careful consideration and the suppliers and manufacturers of plant must be made aware of the potential flow range. Two primary settlement tanks in parallel will provide flow balancing and extra capacity during seasonal peaks; one can then be shut off out of season.

The average sewage organic load from domestic housing is typically 0.060 kg BOD/head/day. Waste disposal units in kitchens significantly increase this load and should be avoided on private systems as they may defeat the aeration system of a small works, generate anaerobic and odorous conditions and increase the amount of settled sludge. Alternatively, the works will have to be increased in capacity by 10–20%.

Grease traps are also essential items for centralised kitchens as small diameter drains and the inlet areas of small works are readily blocked by congealing fat. Flooding, potential pollution, odour and the need for expensive steam jetting are the frequent results. The author has seen a number of such cases at military camps, barracks and conference centres.

Having specified volume and strength and sized the tanks or works accordingly, no indiscriminate additions should be made. Serious hydraulic overloading of small treatment facilities is the most common cause of flooding, treatment failure or effluent non-compliance. It is particularly important that no rainwater drainage is unwittingly connected to the foul drainage and that no surface water enters through poor connections or porous structures.

Equally, there is little point in providing facilities that are too large. Capital and installation costs will be higher and retention times too long in cesspools and septic tanks which may then become odorous. Small treatment works may also suffer from a lack of flow to wet media, an underfed biomass or poor distribution.

Odour and poor quality effluent will often result. If re-circulation of effluent back to the inlet end of the bio-treatment stage is installed to maintain some liquid flow, a lot of power will be used to pump the same water around.

To give some idea of likely plant size, it is usual to allow $1\,m^3$ of filter medium per person in a biofilter, reducing to $0.75\,m^3$ as the plant increases in size to 100 pe.

A biodisc or rotating biological contactor (rbc) expected to produce a 30/20 standard effluent is loaded at $6\,g\,BOD/m^2$ of disc area/day as settled sewage. A 50 pe rbc with an initial settlement chamber will therefore require $50\,m^2$ of disc area, i.e. 25 discs of $1\,m^2$ each side.

The capacity of aeration tanks is based on an allowance of 250 l/head with a retention time of about 18 hours. A 100 pe plant will therefore have a volume of about 18 m^3.

Separate settlement tanks are usually sized to give 6–12 hours retention at average flow rates and therefore tend to be 1/4–1/2 the capacity of the aeration tank.

Air blowers/compressors should be able to supply 17 m^3 of air/head/day to coarse bubble aerators at 2 m depth reducing to 9.5 m^3 at 3.5 m. Fine bubble aeration gives a much more efficient oxygen transfer rate and the above air supply rates can then be halved.

Take the advice of a reputable installer when selecting the plant size. Currently, greywater recycling is being evaluated in parts of the UK. This system uses water from baths, showers and handbasins to flush toilets. After filtration and disinfection, the water is stored in a small tank prior to use. Savings of 5–30% of daily water use per person are reported but the current installation cost of over £1000 generates a long pay-back time in excess of 10 years. For small treatment systems where tankering costs are high or treatment facilities are becoming overloaded by increasing flows, greywater recycling may be well worth considering.

More generally, connection to a private treatment system of fixed capacity dictates sensible water conservation and attention to details and habits. Gravity fed showers use significantly less water than a bath; water saving models of washing machines and dishwashers should be sought; no taps must be left running nor washers and connections dripping.

Another recent development is the Waterfuse, devised by Thames Water plc, which shuts down the mains incoming water supply when demand is excessive or above a pre-set limit.

6.2.2 What level of treatment is achieved or required?

Cesspools are not designed to discharge or effect treatment.

Septic tanks achieve partial treatment by settling out the solids and distributing the effluent through a soakaway where further breakdown occurs in the soil. Permission for installation must be obtained from the Environment Agency who may inspect the discharge if a problem with the tank becomes apparent through local water pollution, ponding or odour.

Small sewage treatment plant are expected to emulate their larger counterparts and achieve 30–60 mg/l Suspended Solids (SS) and 20–40 mg/l Biological Oxygen Demand (BOD). An ammonia (NH3N) standard of 5–10 mg/l may be imposed in sensitive areas. The effluent is thus of sufficient quality to discharge to the local brook or stream although soakaways are not unknown. More relaxed or stringent standards may be required depending on local pollution risk. The majority of contemporary recent package plant achieve better than 30/20 when working properly and some older established filter works generate superb quality effluent. The Environment Agency will issue a Consent to Discharge and may take samples or inspect the plant at random.

6.2.3 What are the installation and site requirements?

6.2.3.1 Cesspools and septic tanks

These are almost always buried and fed by a gravity sewer, a fall of about 1 in 40 being recommended. Installation will require the mechanical excavation of more than the tank volume and problems of sealing may occur where the groundwater table is high. Some careful weighting or partial enclosure in concrete to counter the upthrust of hydrostatic pressure may be needed and septic tanks are usually kept filled with liquid.

They are sited as far from the property as practicable at 15 m minimum, while 25 m is suggested in Building Regulations.

The size of a 3.7 m^3 septic tank for 10 people may be judged from Figure 6.1; a 2.7 m^3 cesspool is in the background.

As most modern tanks are made from grp or plastic based materials, they are not designed to take traffic loads but nevertheless need to be reasonably close to hard standing to provide tanker access.

Septic tanks must be properly ventilated with a soil stack system taking account of prevailing winds and additional ventilation through covers or the drainage system.

An adequate area of ground is required near a septic tank to allow construction of an effective sub-surface soakaway and if some distance from the tank, effluent may have to be pumped to it. A gradient of about 1 in 200 is recommended and the soakaway area should terminate at least 10 m from any watercourse. A series

Figure 6.1 A 10 pe septic tank.

of trenches 300–900 mm wide and at least 500 mm deep are dug usually in a herringbone pattern and porous or perforated pipes laid in them on clinker or gravel. Backfilling with soil over a geotextile membrane will prevent silt entering. Inspection covers at the point of entry and discharge are recommended so that debris and silt can be cleared periodically.

The floor area of the soakaway is a critical design parameter and is covered in depth by British Standard (BS) EN 752-4 1998, while BS 6297 1983 provides details on methods to determine another important parameter, soil porosity or percolation rate. In outline, this involves excavating a hole at the site, 300 mm square and 250 mm below the invert level of the proposed soakaway and filling it with water to a depth of at least 250 mm. This is then left to soak away before refilling the hole to a known water depth and measuring the time in seconds for the water to drain away completely. The experiment should be conducted at least three times in dry weather to obtain an average value for the percolation value in seconds/mm Vp. A satisfactory result is 24 s/mm or less. If Vp exceeds 140 s/mm, BS 6297 1983 considers the soil unsuitable for soakaway construction.

These two factors are then combined in the equation:

$$At = P \times Vp \times 0.25$$

where $At =$ required floor area in m^2 of the soakaway drainage trenches; $P =$ number of persons served by the septic tank; and $Vp =$ soil percolation rate in seconds/mm. Clearly, the greater the value of Vp, the longer the trenches and the greater the area needed. If more than 200 m of trenches are needed or installation of a septic tank and soakaway appears generally difficult, the advice of the Environment Agency and a reputable supplier should be sought and the latter asked to undertake a survey of alternatives. Where there is insufficient land for a soakaway or the top soil is for instance clay with underlying porous chalk, it is usually permissible to drill through and discharge septic tank effluent to the sub-strata.

As mentioned before, permission to discharge any effluent must be obtained from the Environment Agency and Planning Permission and Building Regulation approval is usually required.

6.2.3.2 Sewage works

Many package sewage treatment plant are also buried wholly or in part and similar considerations to those for a septic tank apply. Figure 6.2 shows the installation of a 5 person rotating biological contactor (rbc) treatment plant.

In many cases, the effluent also discharges into the groundwater and the discharge level will require individual setting. Pumping may be necessary both to and from the plant although this should be avoided, as total reliance is then placed on their unfailing operation. The size of some units allows personal entry and as underground plant will constitute a confined space, they must not be entered unattended or before the atmosphere in the chamber has been checked.

Figure 6.2 A 5 pe rbc being installed.

Above ground treatment units are easier to inspect but will be visually apparent unless screened and any odour more noticeable too. Planning permission from the local authority will be required who, since March 1999, has discretion to ask for an Environmental Impact Assessment (EIA).

Careful siting of the treatment plant downwind and some way from housing applies to all installations. Reasonable access for tankers will be necessary during the life of all private systems and for heavy lifting during installation or replacement.

Driveways, roads and aprons should be capable of taking heavy vehicles without damage or jeopardising underground services or structures.

A reliable power supply to most treatment plant is essential as failure of rotation in a rbc or biofilter arm or power loss to aeration equipment in an activated sludge plant may prove expensive to remedy and likely to cause pollution if biological treatment fails. A power failure alarm or rotation failure sensing is essential, particularly where the plant is remote and underground, and the alarm installed where it cannot be ignored!

It should be remembered however that providing gravity flow is possible throughout, an electricity supply is not a pre-requisite for treating sewage. Many small percolating filter works are found in rural areas, some owned by the Water Companies and serving country houses or older estates. These have no 240/415 V supply and any alarms are solar/battery powered. Effluent quality is often excellent and while the capital costs are high, life expectancy exceeds 50 years before major work is required. By contrast, most package treatment plant will last 10–20 years but individual mechanical items of mediocre quality may be life-expired within 5 years.

6.2.4 What are the capital and operating costs of a private treatment system?

The present capital cost of a 9 m^3 cesspool including installation is approximately £4000–£5000 (1999 prices).

A family of four living in a single house connected to this size of cesspool will fill this in about two weeks.

The average cost of emptying this with a standard size 2000 gallon (9 m^3) tanker is £60 (1998). The convenience of 'fit and forget' is heavily off-set by the annual tankering bill, making cesspools an expensive option and ruling in favour of the septic tank if ground conditions permit. However, no sewerage charges will be payable.

The present capital cost of a 4.5 m^3 septic tank including installation and soakaway construction is approximately £3500–£5500 (1999 prices) depending on ground conditions. This would be adequate for 10–14 persons and might require emptying of most of the sludge twice a year at a cost of £60 a visit. However, there are examples of relatively generously sized septic tanks where, due to a large sludge storage volume affording slow but continuous breakdown, the need to de-sludge is rare. Annual operating costs are therefore very low; a life expectancy exceeding 40 years should be expected but the soakaway area may require attention to silting up in less than 10 years.

If additional biological treatment is required for a larger septic tank installation to improve the partially treated effluent, or because a soakaway or borehole is not viable because of ground conditions, capital costs will vary widely depending on process type and are likely to be within the range £10K–£50K for a contributing population of 100. It is assumed land is available without extra cost.

The annual operating costs of separate biological treatment very much depend on type, reliability and sophistication. A gravity fed clinker bed filter of good quality may cost <£10/head/year while mechanical aeration plant with a continuous energy cost and producing sludge after settling for periodic removal may exceed £45/head/year. Effectively, the septic tank becomes the primary settlement section of a small treatment unit but it will often be more economic to buy a purpose-built sewage treatment plant containing all the treatment elements together.

The capital cost of a small sewage treatment works varies widely depending on complexity and level of treatment achieved while its operating costs are often significantly influenced by its energy consumption.

Thus as outlined above, the gravity fed and unpowered percolating filter works at ground level for 200 pe may cost £50K–£200K to build assuming the land was already available but cost <£15/head/year to operate including de-sludging of primary and secondary tanks. A life expectancy exceeding 50 years would be normal.

Table 6.1 Summary of the advantages, disadvantages and costs of cesspools, septic tanks and small treatment works.

Type	Advantages	Disadvantages	Costs	
			Install	Annual operation
Cesspool	Simple; no power, maintenance or equipment	No treatment Regular expense of emptying	£4K–£5K 9 m^3 for 4 pe	£1250
Septic tank	Simple; no power, or equipment	Partial treatment only; soakaway may have limited life and need maintenance	£3.5K–£5.5K 4.5 m^3 for 10–14 pe	£150
Package treatment system (4 pe rbc)	Compact, often underground	Continuous power consumption	£6000	£35–70 per head
	Good effluent quality	Regular maintenance, inspection and periodic de-sludging		
Conventional small STW	V. long life, low running costs; good effluent quality	Visually intrusive, potential odour	£50K–£200K	£15–25 per head
	May not need power	Regular maintenance and de-sludging		

A small underground package rbc plant of 4 person capacity may be obtained and installed for about £6000 and has no significant land requirement. Operating costs/year including electricity would be about £140 or £35/head but a maintenance contract with 2 visits a year and de-sludging will double these values.

The life expectancy before major work or replacement is likely to be 15–20 years. At this point in the life of a lot of package units, complete replacement may be more economic than patching up.

Table 6.1 summarises the advantages, disadvantages and costs of cesspools, septic tanks and small treatment works.

6.3 Commissioning new treatment systems

6.3.1 Cesspools and septic tanks

No commissioning is required with a cesspool as it is only a holding tank. However, after installation and before connecting up to foul drainage, it should be inspected for leaks of surface or ground water over as long a time scale as possible, as these will considerably add to tankering costs.

Septic tanks should be filled with water, tested for leaks and then seeded with digested sludge from a sewage works or sludge from a mature septic tank. About 5% of the tank capacity should be added to speed up the process of maturation and crust development.

6.3.2 Package and small sewage treatment works

Rotating biological contactor (rbc) and percolating filter plants can be seeded with a small quantity of activated sludge from a mature plant mixed with water or diluted sewage.

Activated sludge plants and oxidation ditches should be filled with water to test for leaks and then seeded with about $\frac{1}{3}$ of their volume of activated sludge from a mature plant.

Activated sludge can be obtained from a suitable plant operated by the local water plc. Bucket loads are usually free but a $9\,m^3$ tanker will clearly cost time and travel.

It is imperative that the material is fresh and transported without delay to the new works, where it must be aerated or added to the process immediately.

In all cases, this will greatly assist the process of maturation and allow the treatment unit normally to become fully operational in less than about 6 weeks. Without addition of activated sludge and in cool climates, it can take 6–8 months for a biofilter to mature and produce a good quality effluent and over 3 months for an aerated system to settle down.

6.4 Treatment systems – layout, operation and maintenance

The maintenance requirements and to an extent the operating demands placed upon the owner will reflect the quality of the installation and local factors beyond control – ground conditions, effluent discharge standards and security of power supplies. In the case of package plant, time would be well spent reviewing the pedigree of the supplier and the product and contacting past customers.

There are some areas of self-help however which have been touched upon earlier but are worth re-iteration.

Excessive use or waste of water must be avoided. The annual tankering costs of a cesspool will be greatly affected by this and treatment efficiency in soakaways or complete treatment plants may also decline as a result of hydraulic overloading. Attend without delay to all leaking pipes, sanitary and plumbing fittings.

Grease and fat discharges from cooking must be minimised. At centralised kitchens, staff should be trained to retain fat, grease and chip pan oils for separate disposal or re-cycling. Fat and grease traps should be installed and emptied daily.

All these materials will congeal and separate out in the cooler environment of the sewers and works, block pipes, coat surfaces and weirs and induce rancid odours. Cleaning up is a messy and expensive task and treatment efficiency may be jeopardised. Mineral and heating oils are similarly forbidden materials in the drainage system.

No toxic materials should be discharged to septic tanks or treatment plant. This includes paints, solvents and fuels (which may also generate flammable or explosive conditions in enclosed tanks or package plant), acids and alkalis, solutions containing metals or photographic chemicals and garden pesticides or herbicides. Excessive quantities of cleaners, detergents and bleach will also impair biological action in septic tanks and treatment units, particularly nitrification, and this is often the cause of small package plant failing Consent standards.

Gross solids and slurries will add to tankering costs and the risk of blockages. High organic solids loads or concentrated solutions will tend to overload biological treatment and encourage excessive growth on discs and media, and cause blockages and mechanical overloading. Poor effluent quality may be the result. The installation of sink grinders or waste disposal units to macerate vegetable remnants, meat scraps and greasy material is therefore not recommended.

It is not possible to give hard and fast rules about maintenance and its frequency for package and small sewage treatment plant. Every site is unique as are manufacturers' products, and operating experience will highlight good and indifferent equipment, de-sludging frequencies and failure frequency and effects. Basic ground rules are given below in Sections 6.4.4., 6.4.5 and 6.4.6. However, two tasks are common to all plant – periodic sludge removal and inspection of the operation of the biological oxidation section.

An operating manual or at least a purpose-designed maintenance sheet greatly assists in checking that tasks are being systematically done by a contractor and many suppliers generate their own documentation. A well written manual will allow anyone to operate the treatment plant without prior knowledge, but this pre-supposes that all mechanical plant including valves are tagged, that they are related to a flow diagram of the works and the effect of their operation or failure is known. It also allows a quality audit trail or best practice methods to be followed, a useful defence if the effluent fails Consent standards and the Environment Agency seeks an explanation.

For convenience, package sewage treatment plant are considered separately (in Section 6.4.4) from two examples of smaller versions of conventional plant (Sections 6.4.5 and 6.4.6), as the former are often underground, very compact and maintained as a unit, whilst the latter comprise several sections at ground level requiring individual attention.

6.4.1 Cesspools

Figure 6.3 shows a cross-section through a cesspool. Most installations are now pre-fabricated in glass reinforced plastic although older properties contain

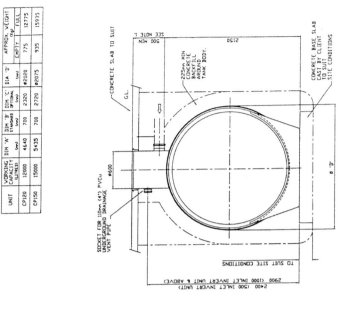
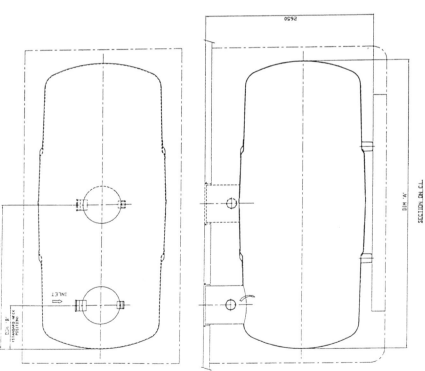

Figure 6.3 Cross-section of a cesspool.

brick-built or concrete structures. The whole tank is buried and connected to the drainage system by either the supplier or the builder.

There are no operational requirements as no treatment is effected.

No maintenance is required other than periodic emptying by a cess-waste tanker. The frequency of this will depend on per capita water use, 180/head/day being a reasonable average. Road tankers generally come in 9 and 18 m^3 sizes.

If the cesspool is not emptied when nearly full, over-flowing at the lowest manhole or from the inspection cover on the cesspool will generate a health hazard and possibly pollute the groundwater or a nearby watercourse.

Sensible water use, attending to leaking valves and washers and a watertight drainage system will ensure economic operation. If the cesspool fills too rapidly, check that its structure is not cracked and admitting groundwater, as may be the case with older brick tanks.

6.4.2 Septic tanks

Figure 6.4 shows a cross-section through a pre-fabricated 3 stage septic tank and representative of many current installations in the 4–40 pe size range.

Earlier brick-built septic tanks for single houses often comprise one square underground chamber with baffled inlet and outlet pipes and sometimes a sloping base at one end to collect the sludge. This layout is shown in Figure 6.5. Two chamber septic tanks are more effective and usually installed for groups of dwellings; most contemporary pre-fabricated tanks have two or more chambers.

A ventilated cover is common to all septic tanks which, with few exceptions, are completely buried to assist in maintaining an even temperature apart from any aesthetic considerations.

There are no operational requirements associated with a septic tank, but it must be remembered that as some biological breakdown through anaerobic digestion of the solids takes place which in turn reduces the frequency for sludge removal and operating cost, care should be taken not to discharge materials toxic to this process as outlined above.

On the maintenance side, most modern septic tanks are of a volume that requires de-sludging every 6–12 months and under-used ones rather longer periods. As an approximate guide, when less than 150 mm exists between the sludge top level and the transfer weir or discharge pipe, it is time to call the tankering service. Ensure that some sludge is left behind during emptying (10–15%), so that sludge digestion can continue with the matured biomass and that the top crust of sludge is not greatly disturbed.

During de-sludging or every 6 months, also check and remove excess floating scum and that the effluent flows freely away. Sludge in the second or third chambers or effluent pipe indicates overloading and the frequency of de-sludging must be increased. If sludge accumulates in the soakaway drains, groundwater pollution and odour nuisance may result and in extreme cases, extensive and expensive soakaway re-building becomes necessary.

166 Sewage and Industrial Effluent Treatment

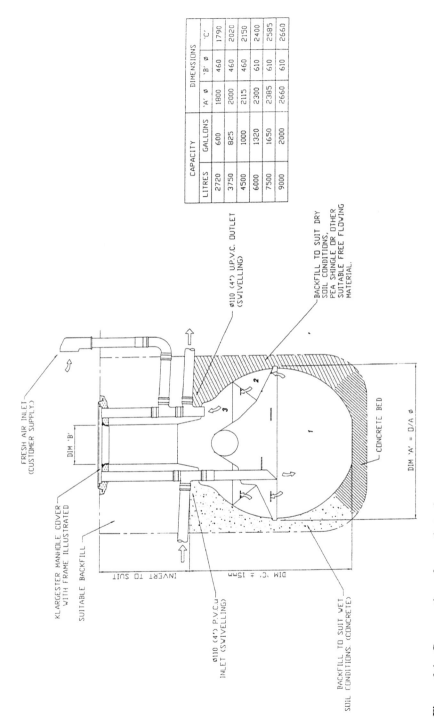

Figure 6.4 Cross-section of a septic tank.

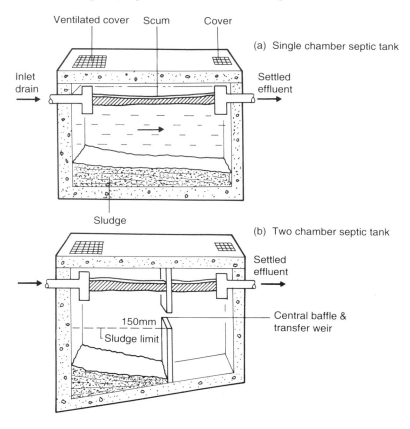

Figure 6.5 Brick built septic tanks.

Finally, periodically inspect and clear ventilation pipes or covers for vegetation; birds' nests are not unknown.

6.4.3 Septic tank soakaways and effluent treatment

The commonest form of effluent disposal is shown in Figure 6.6 and consists of a soakaway area with trenches containing porous pipes laid out in a herring-bone formation. The soakaway area is related to the soil porosity – refer to Section 6.2.3.

Again, there are no operational requirements and maintenance is confined to a 6 monthly periodic inspection of distribution boxes, pipes or channels to clear soil or solids. If the drains become blocked and effluent appears at ground level, odour and insect nuisance is likely and the plot may have to be re-dug or extended.

Impervious clay soils at the surface or less than 1 m below usually rule out the use of a soakaway but discharge to underground porous strata may have been

168 *Sewage and Industrial Effluent Treatment*

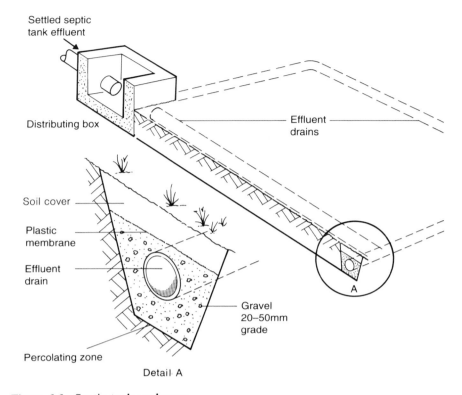

Figure 6.6 Septic tank soakaway.

possible when the septic tank was installed and pipework must be checked every 6 months for blockage.

Where a reed bed or biological filter is installed to give greater effluent treatment, blockage or over-loading with solids must also be avoided. The reader is referred to Section 7.2.3 on reed bed operation and Section 6.4.5 below regarding small biofilter maintenance.

6.4.4 Package sewage treatment plant

6.4.4.1 Layout

Figure 6.7 is a top view of a small biodisc or rbc plant being installed and Figures 6.8 and 6.9 are cross-sections through a simple biodisc or rbc plant and a more sophisticated version with a separate primary settlement tank respectively.

Many package plant are of this type and design; both of these are supplied by Klargester Environmental Engineering Ltd. Further information on rbcs is in Section 4.5.3.

They are designed to achieve the 'conventional' effluent standard of 30 mg/l Suspended Solids (SS) and 20 mg/l Biological Oxygen Demand (BOD). Depend-

Figure 6.7 Small biodisc or rbc plant.

ing on local circumstances, the Consent standards may be more lax than this. More rarely, an ammonia standard of 5–10 mg/l may be imposed and some types of package plant, including rbcs, may require extra oxidation zones to achieve this. Consult the Environment Agency first regarding likely Consent standards and bring these to the attention of potential plant suppliers.

The other common type of package underground plant is a small aeration unit with a single air blower or compressor. Loss of power or aeration will lead to rapid degeneration of the biomass and considerable odour will be generated. A spare blower or compressor is essential for such systems, along with a mains failure alarm.

6.4.4.2 Operation and maintenance

Package plant are normally completely automatic in operation and maintenance tasks will be individual to the type of plant and its size. These must be carefully and regularly carried out to ensure proper operation and treatment as performance guarantees and manufacturers' liability are also usually at stake.

The original suppliers are usually keen to secure the maintenance contract but shop around, as the UK Water plc's also offer this service and sludge tankering facilities.

Electrical power is needed to most types and a power failure alarm is essential. When the mains is restored, any drive motors or pumps should automatically

170 Sewage and Industrial Effluent Treatment

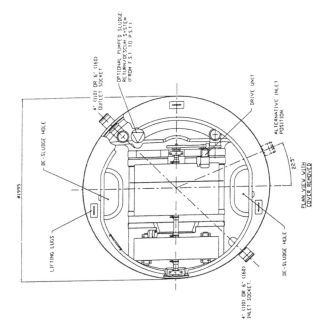

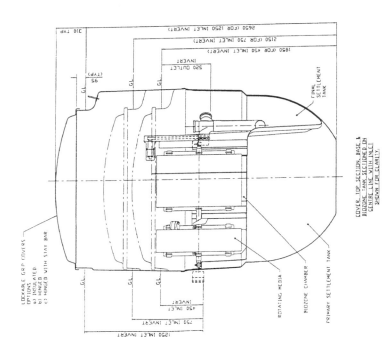

Figure 6.8 Cross-section of a simple rbc.

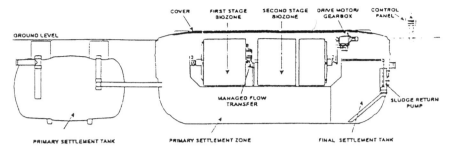

Figure 6.9 Cross-section of a large rbc.

restart. All powered machinery should be regularly tested along with any other alarms – high liquid level, loss of rotation, pump failure, low dissolved oxygen or high levels for instance.

If a power failure occurs for more than 12 hours, uneven growth will occur on the biodiscs in an rbc. The plant must be inspected before re-starting as it may be necessary to scrape the discs clean to ensure even rotation without strain on the transmission.

Power failures for more than 6 hours at aeration plant will induce rapid degeneration and death of the biomass and must be avoided.

A small petrol-driven generator may prove a worthwhile investment if power is lost to the plant with any regularity.

The following gives an indication of maintenance needs for most types of package plant.

Weekly
(1) Inspect air tubes, blowers/air compressors for correct working and delivering air to maintain aeration and mixing. Fine bubble aeration must be maintained by cleaning diffusers and checking pipework for leaks and obstruction. It is strongly recommended that a spare blower or compressor is available or can be quickly made so under any maintenance agreement.
(2) Check all alarm systems.

Monthly
(1) Check the operation of any pumps.
(2) Check the uniformity and thickness of biomass growth on the discs of an rbc and compare with that recommended by the manufacturer. Ensure the discs rotate at a steady speed.
(3) Calibrate oxygen measuring probes – refer to Section 4.5.4.
(4) Check the amount of sludge in an aeration plant.

6–12 monthly
(1) De-sludge underground rbc plant. An aeration plant will probably need some sludge removal at 6 months.
(2) Check the drive chain, grease and re-tension; replace drive belts.
(3) Grease all bearings.
(4) Check the main shaft carrying the discs and covers and supports for corrosion and damage yearly.

6.4.5 A small biofilter sewage treatment works

6.4.5.1 *Layout*

Figure 6.10 shows an example of a single biofilter works serving about 30 houses and laid out like a conventionally sized biofilter works with inlet screen, primary settlement tank, biofilter, two humus tanks and effluent dispersal through a herringbone land treatment area. Most of these units can be seen in the photograph.

Samples on such sites are taken before land treatment because the effluent percolates into the sub-soil. A 30/20 standard is easily achieved with ammonia below 5 mg/l except in the depths of winter when low temperatures inhibit nitrification. Each treatment unit is a miniature version of the full sized plant in operation and the reader is referred to earlier sections of this book for fuller operating details.

Figure 6.10 A single biofilter works.

6.4.5.2 Operation and maintenance

On sites without mains power as shown in Figures 6.10–6.15, flow proceeds by gravity through each stage. Plant operation is therefore automatic and any flow diversion carried out manually during maintenance.

As there is no mechanical equipment at this site, the following maintenance would be typical.

Weekly

(1) Rake the inlet screen to clear detritus. Bag this material and allow it to compost down. As it is a controlled waste, it must not be disposed of with domestic rubbish. Figure 6.11 shows 1 week of accumulation.
(2) De-sludge the humus tanks. If there are two tanks, it may be more convenient to alternate draining down one tank. It may prove possible to de-sludge less often in winter but be aware of rising, de-nitrifying solids that may cause Consent Limit failures; Figure 6.12 shows this occurring on the right hand tank.
(3) Check the biofilter arm rotates easily and the arms or holes are not blocked by rag, solids or grease. Rod the arms out if necessary. Clear leaves and weeds off the media surface.
(4) Check the amount of sludge in the storage bays, Figure 6.13. The frequency of tankering will depend on the amount of drying/dewatering that takes place and odour generation. Spreading raw, untreated sludge to land is now banned in most European countries and the sludge will need digestion or heat treatment at a larger STW before disposal.

Figure 6.11 Manual coarse inlet screen.

Figure 6.12 Humus tanks – one with floating sludge.

Figure 6.13 Sludge storage bays.

Monthly

In addition to the weekly tasks:

(1) De-sludge the primary settlement tank and remove floating scum.
    The tank in Figure 6.14 has not been de-sludged for two months and some septic sludge is apparent.
(2) Check the operation of the distributor in the centre of the biofilter and grease, Figure 6.15; check flow to the arms and rod out or clear the V notches of detritus. Remove any detritus from the media.
(3) Hose down the walls of any tanks or chambers that have been de-sludged or drained and any solids filters or brushes.
(4) Brush tank and overflow weirs and all channels clear of solids or detritus.
(5) Observe the action of the dosing syphon(s). Clean air pipework and check syphon plugs for air tightness afterwards. Wash out the chamber.
(6) Take a sample of the effluent and send for analysis without delay.
(7) Inspect the land treatment area and channels for silting, flooding or vegetation dying back. Divert the flow to other areas as necessary.
(8) Tidy the site and check security.

6–12 monthly

In addition to the weekly and monthly tasks:

(1) Inspect handrailings, steps, grab-irons and walkways for safety and security. Inspect life-saving equipment.
(2) Paint or treat any corrosion on metalwork.

Figure 6.14 Primary settlement tank with floating scum.

Figure 6.15 Biofilter.

(3) Examine wooden scum boards or penstock plates for signs of rotting and replace where necessary.
(4) Re-tension the supporting wires on the biofilter arms.
(5) Examine, and if necessary, rake over the top surface of the biofilter media to remove excessive growths of moss or weeds.
(6) Operate all valves fully and grease spindles.
(7) Check all structures for signs of subsidence or cracking.

Small biofilter works with a mains supply sometimes maintain flow through the filter by using a re-circulation pump and this can improve effluent quality, preventing drying out or freezing. A good quality submersible pump is essential and should be trouble-free for long periods; cheaper ones cannot normally be repaired. The operation of this pump must be checked weekly and the controlling level probes or detector regularly tested. A spare pump should be available at short notice as loss of re-circulation may jeopardise effluent quality.

The satisfactory operation of any other mechanical equipment on a powered site should be checked weekly.

6.4.6 A small activated sludge/oxidation ditch treatment works

6.4.6.1 *Layout*

A plan of an oxidation ditch is shown in Figure 6.16. Otherwise, aeration units are identical (although much smaller) versions of those shown in Figures 4.9 and 4.10 in Sections 4.5.4 and 4.5.5 respectively.

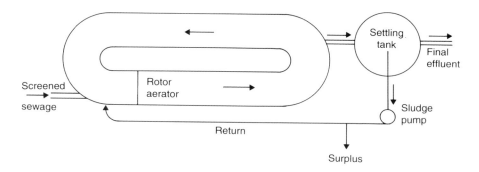

Figure 6.16 Plan of an oxidation ditch.

6.4.6.2 Operation and maintenance

Weekly

(1) Check the efficiency and action of surface aerators or air blowers and diffusers. Check compressor oil levels and traps.
(2) Check all pumps, particularly the return sludge pump from a settlement tank.
(3) Test all alarm systems.
(4) Waste excess sludge from the settlement tank or aeration tank to temporary storage if suspended solids are above the suppliers' recommended levels – usually in the range 2500–6000 mg/l.
(5) Rake clear any manual inlet screen and remove floating detritus from the aeration tank.

Monthly

In addition to the weekly tasks:

(1) Check the settlement characteristics of the sludge using a tall glass jar or cylinder – refer to Section 2.5.3. The sludge should settle quickly leaving a well-defined band between sludge and top liquor. Poor settlement indicates one of several problems – refer to Section 4.5.4.
(2) Tanker excess sludge away; on small works, this is often taken straight from the aeration tank and avoids storage and possible odours.
(3) Clean tank weirs and effluent channels.
(4) Take a sample of the effluent and send for analysis without delay.
(5) Tidy the site and check security.

6–12 monthly

In addition to the weekly and monthly tasks:

(1) Inspect handrailings, steps, grab-irons and walkways for safety and security. Inspect life-saving equipment.
(2) Paint or treat any corrosion on metalwork.

(3) Operate any valves fully and grease spindles.
(4) Check all structures for signs of subsidence or cracking.
(5) Service air blowers/compressors depending on hours run, or as advised by the manufacturers.
(6) On diffused air plant, it will be necessary to remove and clean diffuser domes at 3–5 year intervals. Temporary alternative aeration will have to be installed.

6.5 Conclusions

This chapter has reviewed sewage treatment or containment facilities where connection to main drainage is neither available nor possible.

The majority of good quality plant now available will last many years provided that there is an appreciation of its operation and maintenance and that the responsibility for this ultimately lies with the owner.

6.6 Companies and other organisations

Klargester Environmental Engineering Ltd, College Rd, Aston Clinton, Aylesbury, Bucks HP22 5EW.

Chapter 7
Developments in Wastewater Treatment

7.1 Introduction

This chapter provides a résumé of a number of emerging physical and biological techniques, some of which the author believes will see increased development in the next decade. Several methods that might be described as 'advanced' are also included and, as in many areas of science, some are old techniques in new clothes benefiting from computerised control, plastics and generally better hardware.

For the designer, this overview attempts to provide direction in seeking the 'state of play' of a number of 'new' treatment options, while the plant manager should find it useful when contemplating an upgrade of facilities. It is emphasised that the choice is a personal one, and time and world economics will dictate the course and rate of pollution abatement measures in each country.

An obvious trend is towards the package plant. Sometimes containerised (and therefore transportable) units can be added on quickly where an upgrading of effluent is needed. Larger units tend to be free-standing or at least involve much less civil engineering work and excavation for structures than a classically constructed wastewater treatment plant. Planning to operation times can definitely be 'fast track'.

The oxidation performance of many developments is impressive. Processes are very intensive, removal rates high, retention times low and hydraulic capacity small. This leads to a very compact unit but three aspects are worth highlighting.

(1) The process is very energy intensive in terms of aeration, mixing and perhaps pumping. Power supplies must be very secure.
(2) The small capacity often requires upstream buffering and mixing to iron outflow and influent quality variations. Toxic discharges or extremes of pH will quickly knock the plant out and must be eradicated or a rapid alarm system and flow diversion engineered in. It is particularly noticeable in scientific evaluations subsequently reported in the professional journals that the same brand of package unit often has site-specific performance where flow variation and ambient temperature are important factors; these are often not within the direct control of the operator.

(3) The high quality effluent is produced at a running cost per m^3, cubic metre, which may prove a financial burden over a long period.

The industrial user may also find that some package units are over-engineered in terms of sophistication for their requirements and beyond the abilities of all but the most skilled to repair, or dictate a service contract to maintain them adequately with costly spares on top. Certainly, the treatment plant operator will have to develop skills on a par with those operating the manufacturing process side.

These, then, are some of the basic factors to consider when contemplating a new plant, replacement of present facilities or a 'bolt on'. The processes described are loosely grouped into biological and physical methods although the division is somewhat academic as smaller package plant often contain both.

7.2 Developments in biological treatment

7.2.1 Submerged aerated filters (SAF)

The submerged or flooded aerated biological filter had seen more development in France and abroad than in Britain until a joint venture company, Biwater−OTV Ltd, started to market the patented Biocarbone process in 1990.

In this system, a combination of biological treatment and physical filtration takes place in the same reactor. Settled sewage percolates at a high rate down through a fixed bed of granular material, normally expanded shale in the size range 2−6 mm, produced by firing the natural material.

Air is supplied counter-current via a grid of pipes and diffusers located 20−30 cm above the media base. Figure 6.1 shows a cross-section through a treatment cell made of mild steel and measuring 22 m^2. Media depth is 2 m and volume 44 m^3, the high specific surface (up to 100 m^2/m^3) and the rough porous structure allowing a concentration of biomass four times greater than that in activated sludge plant.

As well as providing an aerated zone, the whole media, particularly the non-aerated section under the diffusers, provides a physical filter for suspended solids. Effluent drains from the base and in order to maintain the filter in a flooded state is conveyed in rising pipework to a washwater storage tank before discharge. The general layout is given in Figure 6.2. There is no separate final settlement tank. Ancillary control, air blowers and pumps are housed in an ISO-sized container. The transportability of the units and the short installation time is a major advantage.

Backwashing is typically every 24 hours and is designed to remove surplus biomass. Duration and airscour rates depend on loading rates, physical characteristics of the medium and biomass growth rates and need setting after

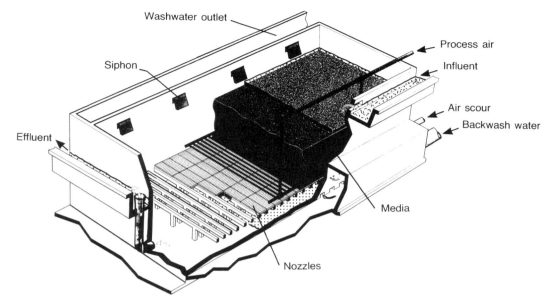

Figure 7.1 Cross-section through a biocarbone cell. (Courtesy: Biwater-OTV Ltd.)

commissioning or pilot studies. Typical backwash liquor solids concentrations are 1000−1800 mg/l and at sewage works installations returned to the works inlet for co-settlement with raw sludge.

Biological performance is high and well reviewed by Lilly *et al.* (1991); a 10:10:5 effluent can be reliably produced at loading rates of 2.5 kg BOD/m^3 media/day, (six times that of conventional activated sludge) and 0.4 kg total nitrogen load/m^3/day. This equates to 85% removals of BOD and SS and over 90% for NH$_3$N.

Power consumption is similar to a nitrifying activated sludge plant at 0.8 kWh/kg BOD and NH$_3$N removed. Hydraulic loading rates up to 100 m^3/m^2 media surface/day are quoted for industrial plant where applications in the USA and Japan include paper mill effluent, oil, milk products and pharmaceuticals. Sewage treatment plant installations have been hydraulically loaded at 20−80 m^3/m^2/day; in France, units are in operation serving populations of 20 000−200 000. Odour and noise are low and the system has been installed enclosed in the basement of buildings.

A recent modification to the Biocarbone system, marketed as Biostyr by Biwater−OTV Ltd, offers denitrification and some degree of phosphorus removal. The main difference is that Biostyr is an upward flow rather than a downward flow biofilter and uses 3 mm expanded polystyrene spheres retained by nozzles at the top of the bed. The regular sphere size has allowed loading 50% higher than with shale media. Air is introduced some way up the filter

182 *Sewage and Industrial Effluent Treatment*

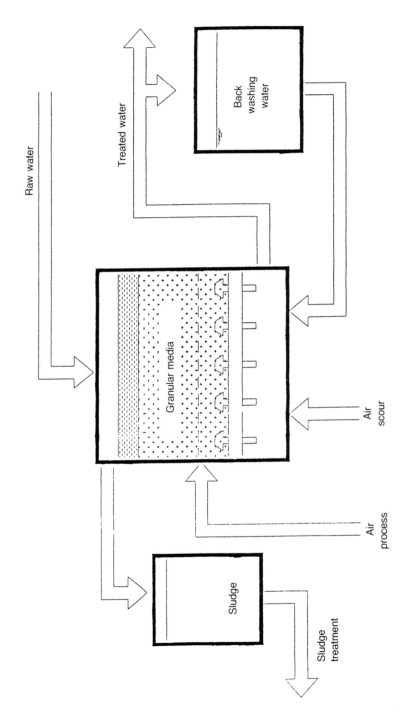

Figure 7.2 Biocarbone process flow diagram.

creating an anoxic zone at the base, and maximum denitrification is encouraged by recirculating some of the effluent.

As the sphere medium floats and is compressed rather than fluidised, oxygen transfer rates are claimed to be higher. Accumulated solids are removed by periodic gravity backwashing from a reservoir of treated effluent. Several sewage treatment plants in France use Biostyr and achieve typical effluent quality of less than 20 mg/l BOD, 15 mg/l SS, total nitrogen of 5–10 mg/l and less than 2 mg/l phosphorus. Where discharges are to designated sensitive waters and nitrogen and phosphorus inputs are limited, this operating variation to the flooded filter will prove a useful alternative.

The UK water industry has mainly used SAF to uprate conventional filter works where space is very limited for extensions, but it has also been used as a nitrifying filter system tacked on the end of a poorly nitrifying plant. Thames Water PLC have developed their own version, the Thames Flooded Filter, into a working design to produce a high quality effluent on sites of limited space; operating experiences are reported by Robinson et al. (1994). Similar in configuration to Biocarbone, this plant has not suffered from the foaming or medium loss seen with other designs. Hydraulic loading has been maintained between 25 and 75 m^3/m^2/day.

The major power cost is for aeration, the volume of air depending on pollution load, endogenous respiration (not significant in an intensive process without the constraints of a sludge age) and oxygen transfer efficiency, which in this design averages 7%. Power consumption has averaged 1 p/m^3 or £50/t of effective oxygen demand (EOD) treated; EOD is defined as the average BOD load plus 4.5 times the NH$_3$N load. Bacterial removals at the plant are reported as a 99.9% reduction of *E. Coli* with an average of 1.6×10^3 in the final effluent – between one and two orders of magnitude better than the conventional biofilters.

A reduction in nitrification has been observed below 8°C, but the effect is less than with an activated sludge plant and a lot less than a percolating filter. Backwash water control has received particular attention on this installation and is returned to the works inlet for co-settlement with primary sludge.

Another version of SAF is marketed by Brightwater Engineering Ltd as Biobead. This has also been used to uprate sewage treatment in the UK and has similar performance to Biocarbone and the Thames Filter. Hydraulic loadings up to 100 m^3/m^2/day are possible, 5 kg BOD/m^3 of media/day representing an upper organic loading rate. The medium is a buoyant, plastic granular material retained by stainless steel mesh and is claimed to have a higher oxygen transfer efficiency, reducing air requirements by two-thirds.

Settled sewage flows upwards through the media bed in the same direction as the air. Cleaning to remove excessive biomass uses air at eight times the normal rate to fluidise the bed, followed by gravity settlement and removal of sludge using the head of water in the unit; no pumps or backwash water and tanks are required. The layout of an operational cell is shown in Figure 6.3.

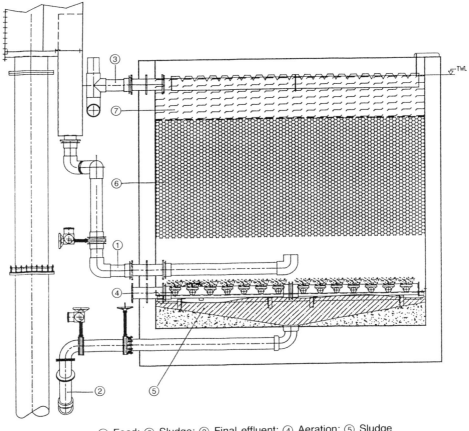

① Feed; ② Sludge; ③ Final effluent; ④ Aeration; ⑤ Sludge settlement zone; ⑥ Media; ⑦ Clarified effluent zone.

Figure 7.3 Cross-section through a biobead cell. (Courtesy: Brightwater Engineering Ltd.)

Both Brightwater Engineering Ltd and WPL Ltd market a package unit incorporating a flooded filter preceded by a primary tank and with a final settling tank, to treat the domestic sewage from small communities up to 175 persons or $35 \, m^3$/day. Figure 6.4 shows the layout: $30 \, SS/20 \, BOD/10 \, NH_3N$ mg/l are the normally achieved discharge standards.

The submerged aerated filter has been demonstrated to provide a compact, effective alternative to activated sludge and percolating filters whilst working at much higher hydraulic loadings. A major advantage is the modular design offering transport to site, minimal civil work and automated operation. Problems of foaming are usually a result of organic underloading. Flow balancing would be advisable in industrial discharges where the flow variation exceeds three times to avoid media and biomass washout.

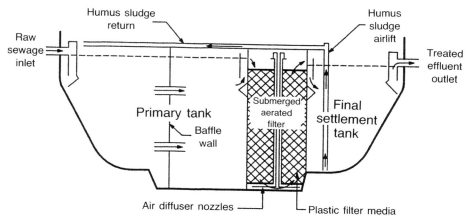

Figure 7.4 Cross-section through a package plant with SAF unit.

7.2.2 Upward flow anaerobic sludge blanket system (UASB)

In its compactness, and being a contained system, this process, developed for the sugar beet processing and related industry in the early 1970s, has similarities to SAF; but operationally it is very different (Figure 6.5) and requires pre- and post-treatment stages either end of the main reaction vessel.

By anaerobic digestion, carbonaceous oxidation of high strength wastewater is achieved with about a 75%–85% COD removal. The system is thus similar to a roughing filter but without the attendant problems of fly nuisance, odour and considerable open space requirements. Influent requirements are a warm, high strength wastewater (COD of 1000 mg/l+) and pH control. Basic pre-treatment after flow balancing is that, in common with anaerobic sludge digestion (see Chapter 5), the acidification vessel is heated to 30–35°C unless the effluent is very warm (this is a primary use for the biogas produced by the reactor).

Other important criteria are:

(1) pH is kept in the range 6.8–7.5.
(2) The COD:N:P ratio is kept at about 350:5:1.
(3) Trace elements are added if not present in the wastewater (Fe, Ni, Se, Mo, Cu, Mn, Cr and Co).

This environment encourages acidifying bacteria to convert sugars to volatile fatty acids (acetic, propionic and butyric). Wastewater then passes to the bottom of the reactor and leaves via V-notch weirs at the top, having been separated by baffles from the methane and carbon dioxide gases produced by anaerobic digestion and the granular sludge. The bottom sludge bed is about 2 m deep and contains thick material (about 10% DM) while the top sludge

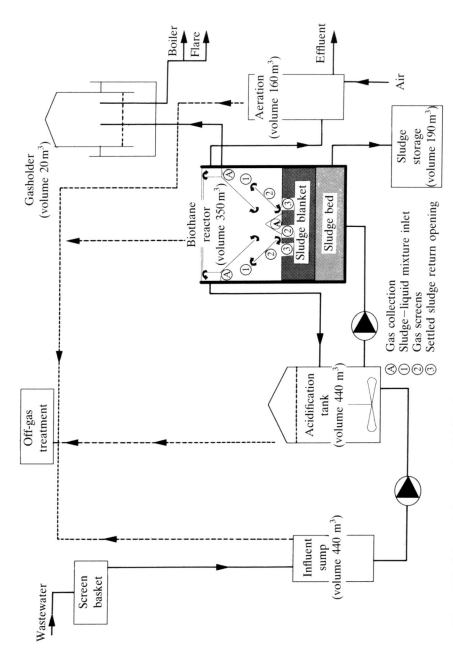

Figure 7.5 Schematic layout of a UASB plant.

blanket is about 2% DM. Figure 6.6 shows the reactor in detail, as supplied by Biwater–OTV Ltd under the name Biopaq.

Two distinct bacterial processes are occurring in the layout seen in Figure 6.5. Acetogenic bacteria convert volatile acids produced in the upstream acidification tank to acetic acid, carbon dioxide and hydrogen which the methanogenic bacteria then convert to carbon dioxide, methane and water.

As the effluent is anaerobic, a post-treatment aeration tank is needed using forced air to strip out hydrogen sulphide. Excess gas can be flared off or used more productively to heat the influent; odorous waste gases are extracted and pass through a peat bed filter.

Typical performance is quoted as up to 90% COD removal at organic loadings of 5–20 kg/COD/m^3/day. Gas production of 0.35 m^3/kg COD removed, retention times of 4–8 hours and a low sludge production of 0.04 kg/kg COD reduced are also characteristic of the process.

The system is marketed as Biothane by Biothane Systems International and has been used worldwide but it is novel in the UK though available through a

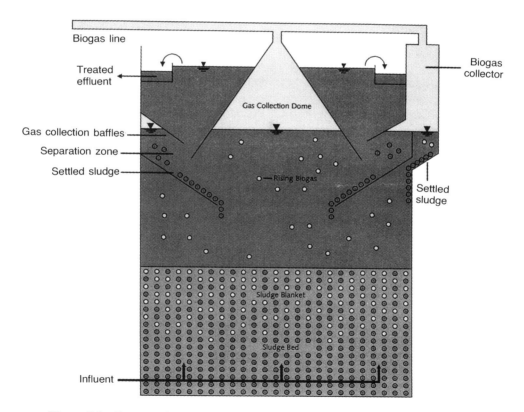

Figure 7.6 Cross-section through a UASB reactor. (Courtesy: Biwater – OTV Ltd.)

licensee (Biwater–OTV Ltd); operating costs are not known. Its recent UK application to the treatment of very strong, sugary effluent from soft drinks manufacture is reported in Housley and Zoutberg (1994), where some difficulties surrounding flow and load balancing are reported. The basic process worked satisfactorily, however; a considerable amount of monitoring via a supervisory control and data acquisition (SCADA) system was installed as the effluent discharged to the foul sewer.

Mosey (1982) describes the application of UASB to sugar beet wastewaters for which it was originally conceived. Wheatley and Cassell (1985) report on extensive pilot-scale trials of UASB using wastewater from a distillery, maltings and a sweet factory, where the importance of pH control and the sensitivity of the bacterial populations to this and shock loads are emphasised.

In view of its comparative novelty and the amount of monitoring and expert control needed, this is arguably not a process for the faint-hearted or where experienced supervision is not readily to hand. Nevertheless, as a compact and entirely enclosed system which can be incorporated into the manufacturing plant on a crowded site and will 'rough' a very strong waste before discharge to the sewer as a trade effluent to reduce effluent disharge costs, it has clear potential as one of few options available at present.

7.2.3 Reed beds

Root zone methods of sewage treatment have seen a recent revival as they offer a virtually maintenance-free method of treating wastewater. Research work in West Germany during the 1970s demonstrated the potential; more recent evaluations have developed the idea to working propositions and several UK Water PLCs have installed reed beds recently. European experience is much wider with over 400 working sites in seven countries.

The reed bed can be used either to treat crude sewage or polish and nitrify secondary effluent and many are started up in the latter mode in low-risk situations. The system also lends itself to treating surface run-off from airports and industrial areas where land is not at a premium and a buffering facility is needed. Almost all schemes screen and degrit sewage before entry to the bed. The bed treats sewage as it flows through soil, furnace ash or gravel in an artificially created wetland containing the reed *Phragmites Australis*.

The hollow rhizomes of the reeds grow horizontally and vertically from growth nodes, generating a hydraulic pathway and treating the sewage by bacterial activity. Oxygen passes from the leaves and stems to the rhizosphere to promote aerobic treatment and then through the rhizomes and out of the root tips. Anoxic and aerobic treatment occurs in the surrounding soil and solids compost aerobically in the above ground layer of dead leaves and stems. Treatment extends to a depth of 300–600 mm. Figure 6.7. shows a typical arrangement.

Primary requirements are level land and a large area; in practice, beds have

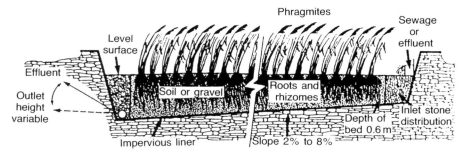

Figure 7.7 Cross-section through a reed bed.

been built with slopes of 2%–8%, the one at Holtby, Yorkshire Water PLC having a 5% slope, operating experiences with this being described by Chalk and Wheale (1989). It has been found that beds that cannot be flooded are often plagued by weeds and the latest recommendation is that slopes should be minimal. Nevertheless the water must pass through the bed at a reasonable rate and UK designs aim for a hydraulic conductivity between 3×10^{-3} and 10^{-4} m/s.

An area of 3–5 m^2/population equivalent is a common yardstick to achieve an effluent with a BOD of less than 20 mg/l on a 95 percentile basis. Most reed beds irrigated with domestic crude sewage will achieve BOD and SS reductions of 70%+ but little nitrification. This is not a disadvantage as many are located on wetland sites and discharge to an estuary with considerable dilution. Phosphorus removal by adsorption is initially high but has proved to be low in UK operating examples. Capital costs of recent schemes have been £50–£150/population equivalent.

A major area of uncertainty and one of continuing research is the growth rate of the reeds. When planted at 2–4 stems/m^2, 2–3 years will elapse in cool temperate climates before the bed is prolific and the use of soil in Germany and Denmark rather than gravel has been found to induce erosion, overland flow and poor reed growth. Some UK installations have suffered badly with weeds and both manual weeding and weedkillers have been employed. In both cases, gravel media is better than soil. Ultimately, the reeds grow strongly enough to smother the weeds, but the ability to flood the bed is useful and terracing has been used on flat sites.

Cooper et al. (1989) provide a useful review of performance during the last decade from a number of reed beds. From this, it is clear that the main advantages of low cost and negligible maintenance are tempered by variable performance caused by variable reed growth rates, weeds and the rooting media. But for small communities or applications where land is plentiful, reed beds provide a viable alternative to conventional treatment plant.

7.2.4 Deep Shaft

Another process developed in the 1970s and arguably therefore not new, this method is the subject of renewed contemporary interest by the UK Water PLCs faced with the problem of providing sewage treatment at coastal sites where visual intrusion must be minimal but a nitrified effluent is not required. The process was developed by ICI for single cell protein manufacture from methanol by aerobic fermentation and is essentially a modified activated sludge plant in a shaft 50–150 m deep and 5–6 m wide. The shaft is fed with screened crude sewage and because of the velocity in the system, grit settlement and ragging do not appear to have been problems.

Figure 6.8 shows a cross-section of the shaft and Figure 6.9 a process diagram of the plant at Tilbury STW, UK, one of the most well-documented installations. Air is injected at a number of points in both the downcomer and riser tubes to aerate and maintain flow round the shaft, the latter being critical to successful circulation and therefore operation.

At the high hydrostatic pressures present in the lower half of the shaft,

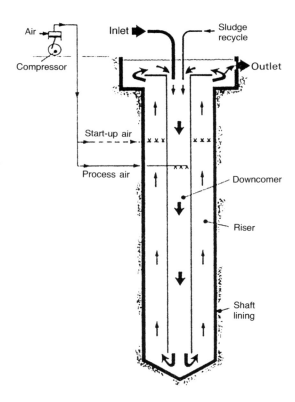

Figure 7.8 Cross-section through a deep shaft. (Courtesy: CIWEM.)

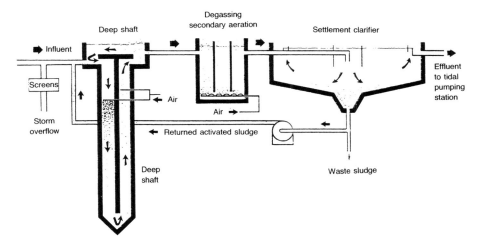

Figure 7.9 Process diagram for a deep shaft. (Courtesy: CIWEM.)

oxygen transfer rates are high, retention time often less than three hours. Sludge ages are therefore never sufficiently long to allow nitrifying bacteria to develop. A degassing tank comprising weirs and gravity settlement is essential at the outlet to remove nitrogen and carbon dioxide prior to recirculation of liquor. The process is best suited to treating high strength waste by carbonaceous oxidation only. Few of the 50 installations worldwide nitrify and many have been used to treat industrial effluent. There are a number of references available describing industrial applications in Europe.

Sewage treatment in the UK has been well documented at the Tilbury plant of Anglian Water PLC where a strong sewage containing much industrial wastewater is treated to a 60 SS/60 BOD standard.

Collins and Elder (1980) and Irwin *et al.* (1989) describe the operation of this plant and modifications that became necessary.

Hemming *et al.* (1977) describe the prototype pilot plant at Billingham STW which atypically treated a normal strength crude sewage of 210 mg/l BOD and 260 mg/l SS. An effluent of 30 SS/20 BOD was regularly achieved with partial nitrification. Operating conditions for Deep Shaft are quoted as:

- Typical BOD and SS removal 90%.
- Typical ammonia removal less than 40%.
- MLSS 2500–6000 mg/l.
- Average sludge age 4 days.
- Sludge production rate 1 kg MLSS/kg BOD removed.
- SVI of sludge 40–100 ml/g.
- F/M ratio range 0.5–1.8.
- Electrical power consumption 0.8–1.0 kWh/kg BOD removed.

Deep Shaft is capable of producing a reasonable quality effluent (60 SS/ 60 BOD) consistently in a very small area with minimal headworks, low odour and the prospect of total automation. Operating experiences indicate that security of power supply to maintain flow round the shaft and effective effluent degassing are the two critical areas. Operating costs are poorly documented but would appear similar to activated sludge plants.

One of the key capital cost areas is the state of the ground in which excavation or drilling will be required. Early plant were installed in old mine shafts and then backfilled, while subsequent upgrades at Tilbury STW were only 60 m deep (not 150 m) after a design evaluation showed this to be the effective minimum depth.

7.2.5 Phosphorus removal

Techniques presently employed for reducing the phosphorus levels in effluents cross the boundary between biological and chemical/physical methods. The two methods used are:

(1) Chemical precipitation of the phosphorus as phosphates using either aluminium or iron salts.
(2) Microbiological absorption in an anaerobic stage of treatment.

Both methods will achieve 80%−90% removals although the general efficiency of phosphorus removal is susceptible to climatic conditions and the nature of the wastewater as outlined below. These factors are relevant to the UK where rainfall frequently becomes a major component of sewage and ambient temperatures tend to be quite low. Sewage contains 5−10 mg/l phosphate and thus 1 mg/l in the effluent is theoretically achievable.

Again, the removal of this nutrient from effluents is not new and has been widely practised in, for example South Africa for a number of years. Barnard (1984) outlines the application of the chemical method of removal by precipitation with ferrous sulphate in a number of 'Bardenpho' plants in that country.

Recent interest in Europe is fired by the requirements of the Urban Wastewater Directive 91/271/EEC (1991) and particularly for discharges to 'sensitive areas' where nutrient input is limited. This will apply to treatment works of more than 10 000 population equivalent and compliance based on 'look-up' tables in a similar way to the existing approach by the NRA.

Chemical dosing can be applied at the works inlet or during final settlement. In an activated sludge plant, chemical dosing with ferrous sulphate is often applied at the aeration plant inlet. Available oxygen oxidises this to ferric phosphate which precipitates in the final tanks; this method is particularly effective in extended aeration (oxidation ditch) plant with long retention times, providing the DO levels are adequate, the anoxic zones not excessively large and MLSS levels below 5000 mg/l.

In a percolating filter plant, chemical dosing is normally applied to the filter effluent before the humus tanks. Thomas and Slaughter (1992) describe operating experiences dosing ferric sulphate at this point at seven STWs in the Broads area of East Anglia, an inland freshwater area that has suffered from eutrophication for 30 years. An increase in sludge production of 15% was reported with an $Fe:PO_4$ ratio of 7:1, achieving 90% reduction of phosphate.

Biological phosphorus removal can be achieved during activated sludge treatment by creating an anaerobic zone prior to the aeration tank. It is essential that fermentation products, e.g. short-chain volatile fatty acids, are present; these are often found in sewage or they can be added as primary raw sludge liquor.

The main responsible organism is *Acinetobacter calcoaceticus*, a strict obligate aerobe, which takes up fatty acids in anaerobic environments (particularly acetic and butyric acid) and stores them intra-cellularly as poly-2-hydroxybutyrate (phb). The energy for this uptake is obtained from the hydrolysis of polyphosphate, stored in the cell, to ortho-phosphate; this is released and the liquid in the anaerobic stage often contains 30−50 mg/l PO_4P.

In the following aerated zone where fatty acids are depleted, *A. calcoaceticus* is highly active, oxidises any remaining carbon sources and reverses the reaction; phb is degraded to acetate and metabolised to produce energy, phosphate being taken up from the liquid and stored again as polyphosphate. The bacterium must then be wasted from the system in the sludge.

Phosphate uptake far exceeds that generated during the anaerobic stage; the more extreme the anaerobic stage and the more volatile acids present, the greater is this 'luxury' phosphate uptake. Research indicates that 7−9 mg/l fatty acids (as acetic acid) are required to remove 1 mg of phosphorus.

Although sewage, particularly if septic, is a ready source of short-chain fatty acids, sewage at temperatures less than 17°C or weak sewage of less than 200 mg/l BOD will contain negligible levels and the anaerobic zone is usually dosed with on-site sources from anaerobic digestor liquor or sludge thickening liquor. These can be high in nitrogen and phosphorus themselves and the favoured source is primary raw sludge left to ferment for a short period. Another option is to dose acetate or acetic acid to the anaerobic stage, the latter as an 80% solution (pure acetic acid freezes at 14−16°C).

Phosphorus removal is often accompanied by nitrogen removal to reduce the eutrophic potential of the effluent to negligible levels and a number of layouts therefore contain an anoxic zone between the anaerobic and aerobic tanks.

Figures 6.10 to 6.15 show the permutations offering both nitrogen and phosphorus removal that have been applied successfully worldwide for some years. The UCT (University of Cape Town) process is marketed in the US as the Virginia Initiative Process (VIP) (Figure 6.14). Cooper *et al.* (1994) describe these individual processes in greater detail and give a good overview

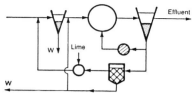

Figure 7.10 Phostrip process for N and P removal (1978).

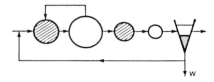

Figure 7.11 Bardenpho process for N and P removal (1974).

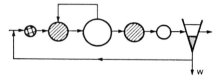

Figure 7.12 Phoredox (Bardenpho modified) process for N and P removal (1976).

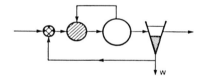

Figure 7.13 Phoredox modified (A^2/O); Bardenpho 3 stage, 1976) process for N and P removal.

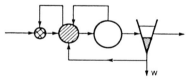

Figure 7.14 UCT (1983), VIP (1987) process for N and P removal.

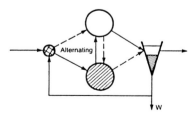

Figure 7.15 Biodenipho (1983) process for N and P removal.

⊗ Anaerobic ⊘ Anoxic ○ Aerobic **W** = Waste sludge

(Courtesy: CIWEM.)

of nitrogen removal methods too. Average percentage removals of phosphorus range between 65% and 95%. Most of these processes can be operated 'side stream' to the main flow and can therefore be used to uprate existing plant without major expense. In some cases, chemical and biological removal methods are operated together, the former being used to support the latter at times of high storm flows when the sewage-derived input of volatile acids declines.

Nutrient removal is here to stay and is likely to become a standard feature of many sewage and industrial effluent treatment plants in the future.

7.3 Developments in physical treatment

7.3.1 Ultra-violet (UV) light irradiation

This disinfection technique, applied to reduce the number of viable bacteria and viruses, has been used to sterilise drinking water for some years. Its use on wastewater as an alternative to chemicals, particularly chlorine, has become established in North America in the last decade with over 700 installations and there is now increasing interest in Europe. One of the motivating aspects is the Bathing Water Directive 76/160/EEC (1976) which places statutory limits of 10 000 total coliforms/100 ml and 2000 faecal coliforms/100 ml.

Sewage works discharges to rivers, estuaries and the sea are point sources of bacterial inputs, as are some treated organic industrial discharges. UV disinfection has been specified or contemplated for a number of STWs in Europe recently where dilution or natural kill rates of pathogens calculated from modelling exercises are insufficient to ensure 100% conformity. The technique has gained acceptance for a number of reasons:

(1) An absence of persistent by-products (particularly compared to chlorine) of uncertain long-term toxicity and with increased mutagenic properties. UV could also generate mutagenic compounds but not at the irradiation levels normally used.
(2) No storage and handling of toxic chemicals. Special operator training, building design and expensive safety requirements are avoided. There is no potential risk to nearby housing.
(3) Reasonable operating and capital costs.
(4) A quick, simple and proven technology.

UV light at 254 nm is readily adsorbed by the nucleic acids in micro-organisms whose maximum absorption of light fortuitously lies in the range 255–260 nm. Damage or rearrangement of the genetic information renders the cell unable to replicate; it is inactivated and dies.

Some reports in the literature indicate that photo-reactivation (self repair) of the microbes occurs in effluent stored in daylight conditions or high in nutrients, but the effects are not consistent and this is a factor for which each effluent would have to be individually tested. Certainly, many micro-organisms have enzymatic repair processes from evolution in naturally solar-irradiated lakes and streams.

The comparative sensitivity of micro-organisms to UV disinfection is listed in Table 6.1. Bacteria and viruses are sensitive to UV at similar doses; one of the widest variations occurs with Rotavirus which is five to ten times more resistant than *E. Coli*. Dose is generally related exponentially to reduction, so doubling the dose gives a 99% (two log) reduction while doubling the dose again (quadrupling the original dose) gives 99.9% (three log) reduction.

A typical installation comprises a number of lamp modules arranged in banks suspended in stainless steel discharge channel(s) with weirs to maintain a constant immersion depth. Easy withdrawal is necessary for periodic cleaning. Switchgear and ballasts are arranged alongside. Lamp failure and low output is sensed automatically and to even out lamp life, duty cycling is common. A 3–4 year lamp-life span is typical. The lamps themselves are low-pressure mercury discharge tubes emitting UV light at 253.7 nm; they are enclosed in quartz sleeves (about 90% UV transmittance) to provide some physical protection. From UK experience, these become coated over in 1–6 months even in good quality sewage effluent by calcium, iron and protein-based

Table 7.1 Comparative sensitivity of bacteria and viruses to UV disinfection.

Organism	Dose in mW/s/cm^2 for a 90% reduction in counts
Bacterium	
Bacillus anthracis	4.5
Bacillus anthracis spores	54.5
Clostridium tetani	12.0
Escherichia coli	3.2
Legionella pneumophila	1.0
Mycobacterium tuberculosis	6.0
Pseudomonas aeruginosa	5.5
Salmonella typhi	2.1
Shigella dysenteriae	2.2
Staphylococcus aureus	5.0
Streptococcus faecalis	4.4
Vibrio comma	6.5
Viruses	
Influenza virus	3.6
Poliovirus	7.5
Rotavirus	11.3
F-specific bacteriophage	6.9

deposits. They can be effectively cleaned by dipping in citric or phosphoric acid. A major supplier of UV equipment is Trojan Industries Incorporated.

In a UV system, the dose delivered depends on:

(1) *The light output of the tube and the equipment configuration.* Doses are measured in milliwatts per square centimetre multiplied by the contact time in seconds, expressed as mW/s/cm^2. Doses applied to wastewaters vary in the range 20–60 mW/s/cm^2. As a guide, to achieve 200 faecal coliforms/100 ml in a high quality effluent of 65% UV transmittance requires 40 mW/s/cm^2

Most UK installations are set to deliver 20–30 mW/s/cm^2 dose rates to effluents that average 50% transmittance and have to achieve better than EC Bathing Water Directive levels. Some safety margin is allowed in dosing calculations and allowance for one order of magnitude regrowth from photo-reactivation is recommended by some researchers.

(2) *The flow-rate of the effluent, operationally controlled to ensure sufficient residence time.*

(3) *The UV transmission (or transmittance) of the effluent.* Sewage effluent contains a cocktail of chemicals, e.g. proteins and urea which absorb UV and reduce light penetration. Typical transmission values are:

- drinking water: 95%
- tertiary effluent: 80%
- secondary effluent, good quality: 65%
- secondary effluent, poor quality: 35%
- storm sewage: 20%
- settled sewage: 5%.

Thus, application to poor quality effluent or partially treated sewage is likely to be expensive in terms of applied dose, electricity costs and the capital costs of lamps, and of variable effectiveness.

The effect of solids on UV dosing rates relates to two factors quite distinct from each other but not greatly influencing transmittance values by more than 1%–5%:

(1) Microbes can be shielded or buried in solids particles.
(2) The size of the solids particles.

Unpublished results indicate inactivation of faecal coliforms in 20 μm particles by 20 mW/s/cm^2 doses, while the average particle size in wastewaters is 30 μm where 30 mW/s/cm^2 is a more typical dose.

Few installations so far have attempted to disinfect any wastewater other than that from a high performance biofilter or activated sludge plant. The performance guarantees recently demanded of suppliers by water utilities have, in the case of UV disinfection, been dependent on performance of preceding activated sludge and settlement units and suspended solids/

transmittance values of the effluent. It seems likely that future NRA Discharge Consents with bacterial standards will dictate UV dosing levels related to effluent quality, rather than the measurement of bacterial numbers and therefore cover how the plant is operated.

Interesting accounts of the UV disinfection of sewage works effluent (at the Bellozanne plant in Jersey, Channel Islands) have been described by Gross and Davis (1991) and Gross and Murphy (1993). A general consensus of costs indicate that UV can, at best, halve the operating cost of chlorine disinfection, with electricity costs, effluent quality and lamp life being the major factors. Capital costs are similar to chemical dosing systems.

7.3.2 Wet oxidation processes

A number of chemicals present in low concentrations in industrial effluents are resistant to normal biological methods of breakdown and often highly toxic to treatment processes; they also generate by-products which undergo very slow degradation in the natural environment. As a result, these chemicals end up in food chains, accumulate in fat tissue and present taste and odour problems in water supplies. The main culprits are the organo-halide solvents, e.g. tri- and tetra-chloroethylene, chloro-phenols, nitro-phenol dye wastes and other aliphatic and aromatic organics.

Many of these are now being severely restricted or banned by some countries and limits in effluents are likely to become very tight for individual compounds, with the added penalty of higher charges for the 'organic strength' factor to which they contribute.

Most oxidation processes centre on the use of hydrogen peroxide or ozone, with the use of a catalyst or UV light. Low temperatures (100°C) and pressure are also employed to generate free radicals (usually OH^-) whose reaction rate is 10^6-10^9 faster than that of strong oxidants such as ozone. Both solids and liquids are amenable providing the aqueous phase is freely available chemically, although most reactors are operated with sludges of more than 30% DM. The end products are usually CO_2, H_2O and a mixture of salts or acids. Even if complete mineralisation is impractical, oxidation processes still offer a useful pretreatment and 80%–90% reductions in the 'strength' determinands of TOC and COD.

A safety factor to note is the possible generation of oxygen from H_2O_2 decomposition and the explosion risk with solvent/oxygen mixtures in the headspace of the reactor.

The general formula for the reaction of organic waste with hydrogen peroxide and a catalyst at 100°C is:

$$CxHyOz\ (X) + (2x + 1/2y - z)\ H_2O_2 = x\ CO_2 + (2x + y - z)\ H_2O + (X)$$

where X = a functional group

Breakdown products tend to be polar and remain in solution until degraded to carbon dioxide and water; low molecular weight solvents will volatilise. Reaction efficiency is measured by the proportion of CO_2 in the gases emitted, more than 90% being achievable. The technique has been developed in the UK to treat radioactive wastes by the Process Engineering Research Centre, and for more general commercial use by Leigh Environmental Ltd.

Photolytic oxidation by UV light coupled with a strong oxidant like hydrogen peroxide or ozone is the subject of current research work at Bath and Bradford Universities in the UK. UV with a photocatalyst (semi-conductive titanium dioxide) is also being studied.

While it is clear from research papers that advanced oxidation techniques are still in their comparative infancy, they will certainly be developed in the next decade as a rival to incineration, to handle the small volumes of intractable wastewaters and sludges for which a declining number of disposal outlets are available.

7.3.3 Lamella plate separators

These offer a method of increasing the solids removal performance of a plant without resorting to building more settlement tanks. In many industrial situations where space is at a premium, the ability to 'drop in' an upgrade is always potentially useful. Similarly, in the municipal sewage treatment plant the compactness offered by lamella plates has encouraged those responsible for some estuarine and coastal treatment schemes to look at underground installation below car parks, sports grounds and cliffs. Their use is common in Europe; in France, a number have been operating for up to five years in combination with SAF plants (see Section 6.2.1). Little information on current performance, reliability and operating costs is yet available.

Past experience in the UK is largely anecdotal and not widely documented. The most successful application to date seems to have been in humus tanks to enhance solids removal to achieve Consent standards. Historically, lamella inclined plates have also been used for oil/water separation and there are a number of vintage references to this application. No current installations in the UK are documented. It is understood that several Water PLCs are following the continental lead and constructing submerged aerated filter plants (SAF) in which lamella plates are used to provide enhanced primary sedimentation in a confined space.

Each lamella plate unit consists of corrugated inclined sheets held in a frame and normally constructed in stainless steel. Flow enters the plate area from either side and passes upwards, exiting via distribution orifices into the effluent discharge flume.

Solids collect on the inclined plates and roll downwards, counter-current to the upward flow, falling to the tank base for removal. The plate inclination is normally 45°–60° above the horizontal and represents a compromise between solids accumulation and reduced efficiency. A cross-section of the flow patterns in an installation supplied by PWT Projects Ltd is shown in Figure 6.16. As the plates are 'stacked', the effective settling area of each unit equals the area of each inclined plate projected onto a horizontal surface, multiplied by the number of plates in the unit – usually ten. Thus the projected solids removal area is at least ten times the surface area of the tank containing the lamella plates.

As an example, at an installation in Stockholm for secondary settlement after biological treatment, the tanks occupy $200\,m^2$, but have $5000\,m^2$ of settlement area. Primary treatment plant have typically 6 mm fine screens before the plates to avoid blockages by rag and detritus. General operating conditions and performance are as follows:

- Upward flow rate: 5–25 m/h.
- Operation without chemicals possible at flow rates less than 8 m/h.
- Coagulant dose (ferric chloride or sulphate): 15–40 mg/l.
- Flocculant dose (polymer): 0.5–1 mg/l.
- Typical SS and BOD removals: 40%–75%.

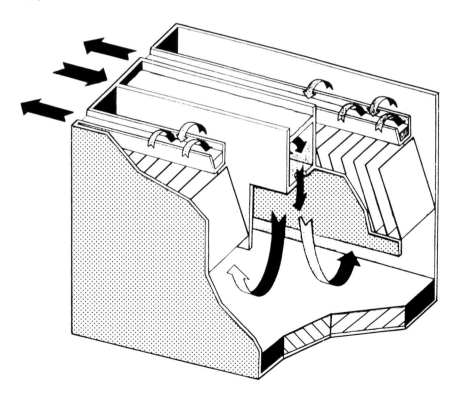

Figure 7.16 Flow patterns through a lamella plate separator.

- Typical phosphorus removals: 50–60%.
- Typical sludge: 5%–8% DM (3%–5% without chemicals).

Critical to success are the upward flow rate, maintenance of fairly stable flow distribution conditions and regular sludge withdrawal. Merely suspending a plate pack in a settlement tank is unlikely to be productive unless these aspects are checked first. Some preliminary laboratory tests to assess the rolling ability of the type of solids to be collected are also essential.

Typical maintenance of lamella tubes in sewage treatment consists of cleaning every month in summer, once every two months in winter. The tank is partially drained down and the plates pressure-cleaned with good quality final effluent. Some wastewaters will encourage more rapid clogging by biological growth or a build-up of grease and oil. Operating costs will be much influenced by bulk chemical costs and dose rates; few lamella plate installations are operated without chemical dosing.

An aspect relevant to the use of lamella plates prior to SAF is the removal of nutrient phosphorus, but the limited reviews so far do not indicate any operating problems. It should be remembered that the basis for setting effluent Consent standards varies widely within Europe.

The treatability of sludges from lamella plates using iron salts and low doses of polyelectrolyte is an important side issue. Black, odorous sludges up to 10% DM are reported from a pilot plant, from which a cake of 30%+ DM could be obtained. Filtrate quality is likely to be poor and odour problems might be expected on a full-scale plant. Enhanced settlement by lamella plates offers much promise for solids removal where space is at a premium, for 90% savings in land area are possible and lamella plates can be more efficient than most conventional gravity settlement methods.

7.3.4 Membrane filtration

The improvement in manufactured quality and reducing cost of semi-permeable and fabric membranes with particle size retentions below 1 μm have encouraged experiments and the manufacture of packaged plant designed to treat sewage and selectively remove trace organics. The use of membranes for selective organics removal permits the reclamation of valuable product at both production and laboratory level. Other important applications include potable water boreholes and the desalination of seawater by reverse osmosis.

Membrane filtration of wastewater, tried successfully in Australia and the USA is beginning to gain ground in Europe as a viable method. Variously described as cross-flow, ultra or microfiltration depending on the *modus operandi*, removal of finely divided colloidal material, associated BOD and COD and considerable reductions in bacterial and viral numbers have been shown to be possible.

For sewage treatment, the technology offers a complete break with conventional sedimentation tanks and disinfection by chemicals or UV. Several plant have been installed in coastal regions of the UK to provide primary treatment as required by the Bathing Water Directive 76/160 EEC (1976) and the Urban Wastewater Directive 91/271/EEC (1991). The primary reason for its application has been to reduce bacterial and viral numbers to conform with the former Directive.

The construction material of most commercial membrane packs are tubes made from polysulphone or polypropylene fibres, both chemically inert and temperature-resistant. Ultrafiltration membranes have pore sizes below $0.01\,\mu$m and operate at high pressure; microfiltration membranes operate at much lower pressures of $30-70\,$kPa with pore sizes above $0.05\,\mu$m.

Commercial membrane installations are modular, each unit containing over $50\,\text{m}^2$ of membrane area. Life expectancy is application dependent but the manufacturers predict 3–5 years. Running and replacement costs at present roughly equate to conventional treatment but it is predicted these will drop substantially as the technology is developed.

Effective settled sewage treatment by microfiltration relies on an initial build-up of solids on the membrane surface to form a filtering medium. Flow then passes down the tubes at a rate which maintains debris in suspension and has a mild stripping action on the deposited solids. In time, the solids layer becomes excessive and mechanical or compressed air cleaning of the tubes is necessary. The process is thus a batch one unless many units are operating in parallel, requiring header tanks and preliminary chemical dosing; aluminium sulphate is commonly used.

Figure 6.17 shows the layout of a 4500 population equivalent plant installed for Welsh Water PLC by PWT Projects Ltd; two microfiltration units which provide effective barrier filtration to $0.2\,\mu$m treat settled sewage from lamella plate separators prior to discharge to sea. General performance figures of 95% SS, 68% COD and 65% BOD removals are achievable, with a 10^3 reduction of indicator bacteria. Results from the USA and Australia, where membrane filtration is more widely used, indicate virtually 100% removals of viruses and faecal bacteria. This has led to their application as a terminal disinfection technique in these countries.

Another application is the treatment of oily wastes from machining, food manufacture and ships bilges, where emulsions are often formed. Typical oil droplet size is $0.1-0.5\,\mu$m and microfiltration hydrophilic membranes will work well at low pressures, the membrane surface encouraging a static water layer to form and prevent oil breakthrough. This is likely to become an important method as Consent limits as low as 10 mg/l are set for oil and emulsions in Europe.

The concentration of surplus activated and anaerobically digested sludges by cross-flow microfiltration is reported by Bindoff *et al.* (1988).

Developments in Wastewater Treatment 203

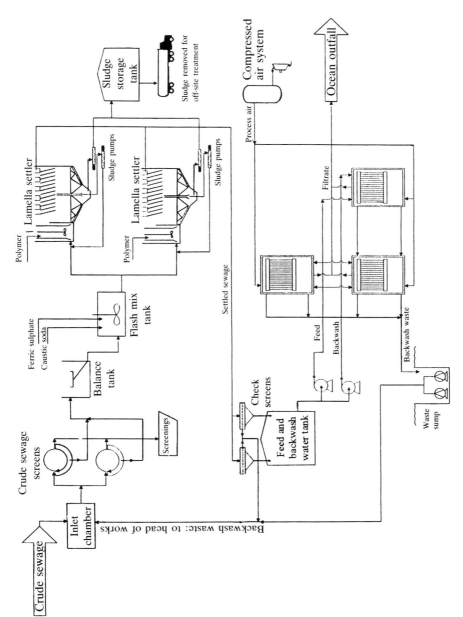

Figure 7.17 Schematic layout of a microfiltration STW. (Courtesy: Memcor Ltd.)

7.3.5 Genetic engineering

It is appropriate to conclude this chapter with a brief account of the present 'state of play' of an emerging method which may have considerable impact on the whole philosophy and practicality of wastewater treatment. Genetic engineering, the insertion of nucleic acid molecules artificially to form new combinations of heritable material, has already been applied to agriculture and food production. The primary purposes here have been to increase production output and environmental resistance to attack from competing organisms.

Bioremediation, the application of indigenous micro-organisms to enhance wastewater treatment, has been applied successfully to treat oil spillages and 'kick-start' anaerobic digestion and composting.

However, much of the potential for applying genetic engineering to waste treatment has yet to be realised. Present interest lies in the genetic modification of bacteria to enhance the breakdown of recalcitrant and toxic compounds, the UK 'Red List' substances providing a focus (Table 6.2). Halogenated

Table 7.2 UK Red List and Black List substances.

Red List	Black List (additions to Red List)
Mercury and compounds	2-Amino-4-chlorophenol
Cadmium and compounds	Anthracene
Gamma hexachlorocyclohexane	Azinphos-ethyl
DDT	Biphenyl
Pentachlorphenol	Chloroacetic acid
Hexachlorobenzene	2-Chloroethanol
Hexachlorobutadiene	4-Chloro-2-nitrotoluene
Aldrin	Cyanuric chloride
Dieldrin	2-4-D
Endrin	Demeton
PCBs	1,4-Dichlorobenzene
Dichlorvos	1,1-Dichlorethylene
1,2-Dichloroethane	1,3-Dichloropropan-2-ol
Trichlorobenzene	1,3-Dichloropropene
Atrazine	Dimethoate
Simazine	Ethylbenzene
Tributyl and triphenyltin compounds	Tenthion
Trifluralin	Hexachloroethane
Fenitrothion	Linuron
Azinphos-methyl	Mevinphos
Malathion	Parathion
Endosulfan	Carbon tetrachloride
	Chloroform
	1,1,1-Trichlorethane
	Chloridazon
	Chloroprene

aromatic hydrocarbons are a particular area of interest, being potentially major global pollutants and xenobiotic, i.e. not normally formed by natural biochemical processes.

In experiments with a laboratory scale activated sludge unit (see McClure et al., 1991) a genetically engineered strain of Pseudomonas was introduced for the specific purpose of catabolising 3-chlorobenzoate. No breakdown was observed and this was attributed to low numbers of viable organisms, the low temperature of the liquor (30°C would be optimum) and the presence of alternative and more attractive substrates. Continuation of the experiments showed that selective adaption could be encouraged in naturally-occurring bacteria cultured in a 3-chlorobenzoate enriched medium. Further strain selection and manipulation of inoculant density would probably lead to better biodegradation rates.

This might be an attractive alternative strategy. The application of strong selective pressures to generate enriched inocula of indigenous bacteria and more likely to survive the competition in a heavily contaminated environment, also circumnavigates pending legislation on the release of genetically engineered organisms to the environment. This is the political response to public concern and debate in an emotive area of science. Time will tell how research continues and what direction it takes.

7.4 References

Barnard, J.L. (1984) Design and operation of Bardenpho plants in an African country. *Journal of the Institution of Water Pollution Control*, **83** (4).

Bindoff, A.M., Treffry-Goatley, K., Fortmann, N.E., Hunt, J.W. & Buckley, C.A. (1988) The application of cross-flow microfiltration technology to the concentration of sewage works sludge streams. *Journal of the Institution of Water and Environmental Management*, **2** (5).

Chalk, E. & Wheale, G. (1989) The root zone process at Holtby Sewage Treatment Works. *Journal of the Institution of Water and Environmental Management*, **3** (2).

Collins, O.C. & Elder, M.D. (1980) Experience in operating the Deep shaft activated sludge process. *Journal of the Institution of Water Pollution Control*, **79** (2).

Cooper, P., Day, M. & Thomas, V. (1994) Process options for phosphorus and nitrogen removal from wastewater. *Journal of the Institution of Water and Environmental Management*, **8** (1).

Cooper, P.F., Hobson, J.A. & Jones, S. (1989) Sewage treatment by reed bed systems. *Journal of the Institution of Water and Environmental Management*, **3** (1).

Directive 91/271 EEC (1991) Council Directive Concerning Urban Wastewater Treatment. *Official Journal L135/40*.

Directive 76/160 EEC (1976) Council Directive Concerning the Quality of Bathing Water. *Official Journal L31/1*.

Gross, T.S.C. & Davis, M.K. (1991) Ultra-violet disinfection of sewage effluents: a pilot study at Bellozanne, Jersey. *Journal of the Institution of Water and Environmental Management*, **5** (6).

Gross, T.S.C. & Murphy, R. (1993) Disinfection of sewage effluents: the Jersey experience. *Journal of the Institution of Water and Environmental Management*, **7** (5).

Hemming, M.L., Ousby, J.C., Plowwright, D.R. & Walker, J. (1977) Deep shaft – latest position. *Journal of the Institution of Water Pollution Control*, **76** (4).

Housley, J.N. & Zoutberg, G.R. (1994) Application of the 'Biothane' wastewater treatment system in the soft drinks industry. *Journal of the Institution of Water and Environmental Management*, **8** (3).

Irwin, R.A., Brignal, W.J. & Biss, M.A. (1989) Experiences with the Deepshapft process at Tilbury. *Journal of the Institution of Water and Environmental Management*, **3** (3).

Lilly, W., Bourn, G., Crabtree, H., Upton, J. & Thomas, V. (1991) The production of high-quality effluents in sewage treatment using the biocarbone process. *Journal of the Institution of Water and Environmental Management*, **5** (2).

McClure, N.C., Fry, J.C. & Weightman, A.J. (1991) Genetic engineering for wastewater treatment. *Journal of the Institution of Water and Environmental Management*, **5** (6).

Mosey, F.E. (1982) New developments in the anaerobic treatment of industrial trial wastes. *Journal of the Institution of Water Pollution Control*, **81** (4).

Robinson, A.B., Brignal, W.J. & Smith, A.J. (1994) Construction and operation of a submerged aerated filter sewage treatment works. *Journal of the Institution of Water and Enviromental Management*, **8** (2).

Thomas, C. & Slaughter, R. (1992) Phosphate reduction in sewage effluents: some practical experiences. *Journal of the Institution of Water and Environmental Management*, **6** (2).

Wheatley, A.D. & Cassell, L. (1985) Effluent treatment by anaerobic biofiltration. *Journal of the Institution of Water Pollution Control*, **84** (1).

7.5 Companies and other organisations

AEA Technology, Effluent Processing Club, 404, Harwell, Didcot, Oxfordshire OX11 ORA.
Biothane Systems International, Delft, The Netherlands.
Biwater–OTV Ltd, Biwater Place, Gregge Street, Heywood, Lancashire OL10 2DX.
Brightwater Engineering Ltd, Centre 64, Wilbury Way, Hitchin, Hertfordshire SG4 OTP.
Leigh Environmental Ltd, Lindon Road, Brownhills, Walsall, West Midlands WS8 7BB.
Memcor Ltd, Wirksworth, Derbyshire DE4 4BG.
PWT Projects Ltd, 632/652 London Road, Isleworth, Middlesex TW7 4EZ,
Sunwater Ltd, 44 Friar Street, Droitwich, Worcestershire WR9 8ED. (Sole UK agents for Trojan Industries Inc, 845 Consortium Court, London, Ontario. Canada.)
WPL Ltd, 14/15 Bridge Industries, Broadcut, Fareham, Hampshire PO16 8SX.

Chapter 8
Sampling and Analysis

8.1 Introduction

Adequate sampling of the effluent from a factory, sewage works or the intermediate stages of any treatment plant, coupled with accurate, relevant analysis are essential to maintain efficient operation, not only of the treatment plant, but often also of the factory itself. Poor quality effluent often indicates a poorly maintained or incorrectly loaded treatment system which is probably failing Discharge Consent limits and frequently points to equally poor quality control, waste of materials and water, and process malfunction within the manufacturing area.

In the UK and Europe, there is a trend in recent legislation for companies and statutory water undertakings to be required to monitor and keep publicly accessible records about the nature and quality of effluent discharges from treatment works. And yet, sampling and analysis are two areas of wastewater treatment which are frequently under-resourced. Many industrial sites rarely sample their own effluents, and certainly never monitor intermediate stages, unless provoked to do so by the regulatory body who have just collected a sample clearly outside the limits. Heavy reliance is placed upon the subjective assessment of an operator, whose main function is often mechanical, making up dosing chemicals and repairing pumps. The walk around the factory site and treatment plant by the site supervisor, and confined to a pleasant sunny afternoon, rarely seems to catch the rising sludge, loss of pH control and the black effluent.

Some sites build up an impressive dossier of effluent results confined to one minute in 24 hours, Monday to Friday; 9.30 AM is a popular time. This provides a near-definitive snapshot of the situation near to the morning tea-break, but reveals nothing about the other 99.93% of the 24-hour period.

By one of the ancient laws governing human behaviour, the effluent inspector regularly turns up to take a sample five minutes after the discharge of gin-clear liquid has ceased, and the factory or plant is discharging something resembling brown soup. This scenario can be expensive on three counts:

(1) Fines imposed for breaching consent limits are much heavier than ten years ago in many countries, particularly within the EU and the USA.

(2) The unattractive publicity attached to polluting a river always carries a negative sales and public relations tag which rapidly travels worldwide.
(3) The charging formula used by many authorities contains at least one element relating to effluent strength, often measured as Chemical Oxygen Demand (COD). One sample of high values in a 6- or 12-month charging period often skews the average of the other ten or so samples that are taken. An unexpectedly large bill for effluent charges is received. The charging formula and some of its implications presently applicable to the UK and similar to many others worldwide is discussed later in this chapter.

So it is important to design a routine sampling programme in line with the complexity and strategic importance of the site; resource the programme adequately ensuring the provision of analytical facilities and stick to it; act on the results and use them to tune the plant or make production changes and economies.

Equally important is relevant, good quality analysis. Many larger factories will have their own laboratories devoted to quality control analysis of raw materials and product. With a little commitment in time and money, it is quite possible to carry out a few simple and fairly quick tests on the effluent to check whether it conforms to discharge standards. Specific methods used by the water industry will be found in the well-documented and thoroughly recommended collection issued by HMSO/Standing Committee of Analysts in the UK.

Many other countries have similar standard wastewater analysis methods and the water supply undertaking or institution representing scientists and engineers in this field should be able to advise.

This chapter also reviews contemporary methods for flow measurement and on-line sensors and electrodes that will give continuous, good quality readings for a variety of parameters in what are often hostile and physically difficult environments.

8.2 Sampling: initial considerations

In planning a sampling scheme, the objectives have firstly to be defined. Often there is more than one objective and the list will include some of the following:

(1) determining the concentration or pollutant load being discharged;
(2) providing operational data for plant personnel;
(3) testing compliance with discharge load or concentration limits;
(4) providing data for charging purposes or the maintenance of a legally required public register.

Most wastewater treatment plant do not perform consistently because the received influent is of varying quality and flow rates are rarely constant. The process is in any case mainly biological and will be influenced by external factors, particularly ambient temperature. So the best overall picture will be obtained by taking and analysing the highest number of samples compatible with resources over a time period likely to capture most variations of flow rate, influent and effluent quality and routine operating procedures.

On many sites, automatic samplers that can be connected to flow recorders or take a number of samples at timed intervals with or without bulking them into a composite will provide the answer; they also provide an invaluable record of plant performance during unmanned periods. Automatic samplers are discussed in greater detail below.

8.3 Sampling methods

8.3.1 Continuous

A possible solution to the limited sampling resources is continuous on-line monitoring by automatic equipment that also analyses the liquid for determinands of interest immediately. This is particularly important if the wastewater is organically polluted and the components are liable to biodegrade quickly, affecting the values in subsequent analysis. For discharges containing inorganic materials, on-line analysis provides a near-continuous check of the discharge which may have important control or alarm functions on manufacturing processes, but is otherwise not essential.

To obtain a value for the polluting strength of the liquid rapidly, the current generation of analysers concentrates on measuring Total organic carbon (TOC) by either high temperature catalytic oxidation or ultra-violet (UV) irradiation of the sample, followed by measurement of the carbon dioxide released. Detection limits are usually better than 10 mg/l and reaction time 10 minutes.

The equipment takes a discrete sample at short intervals, and in that sense is not a continuous monitor, but it must be realised that in order to sample continuously most effluents representatively would dictate an abstraction rate sufficient to keep solids in suspension and this would practically mean taking $2-3 \, m^3$/day, an impossibly large volume for analysis.

In its disfavour, such equipment is often expensive and limited in performance, both in terms of analytical accuracy and range of determinands that can be analysed. Volatile materials are difficult to measure as most equipment sparges the sample with nitrogen to remove carbonates and will purge volatiles out. Non-carbonaceous compounds that contribute to oxygen demand will also be missed, e.g. amines. An estimated compensation for inorganic carbon compounds in the effluent has also to be made.

By its nature wastewater (often high in solids, grease and materials which will block or corrode small pipes in such equipment) provides an aggressive environment for on-line monitors, and much time can be spent checking, cleaning and recalibrating such systems. They are, however, useful on remote sites where data may be transmitted automatically by telemetry, attendance is infrequent or 24 hour surveillance is needed.

At large complexes on-line monitoring has its place measuring TOC to give an indication of overall organic load being discharged.

Among suppliers in the UK are Ionics UK Ltd and Pollution Prevention Monitoring Ltd.

8.3.2 Flow proportional

Of the various methods of sampling involving the taking of discrete samples at intervals, flow-proportional sampling is the most accurate; it is indeed demanded by law in some EU countries. In the UK, Section 121 Paragraph 2 of the Water Industry Act 1991 outlines conditions that may be attached to a trade effluent Consent requiring the discharger to provide and maintain meters to measure the volume and rate of discharge and apparatus to determine 'nature and composition of any trade effluent'. In recent times, there has been increasing enforcement of these requirements and organic dischargers originally on fixed and very low charges have had unpleasant surprises as to the true composition of their effluents.

However, as Table 7.1 illustrates, flow-proportional can work much in the favour of a factory used to spot sampling once a day at three different times. In this example, no effluent was discharged at night, the daily flow being 1395 m^3 and average discharge rate being 10 l/s, with peaks during wash-ups of 20–30 l/s at 1000 hours and 1630 hours. The average COD and SS were 600 and 400 mg/l respectively. The hit and miss nature of spot sampling is highlighted, and the wide variation in charges likely from this method.

Equipment comprises an ultrasonic detector measuring liquid height as it flows through an accurately constructed flume or over a weir. A program within the controller uses the variation in liquid height to calculate flow once the dimensions of the weir or flume have been entered during setting up. A

Table 8.1 Table of sampling times/analysis/costs. (Courtesy: J. Corris, Warren Jones Engineering.)

Spot samples	COD	SS	Cost (£) per annum (240 days)
1000 hours	600	400	91 829
1300 hours	400	267	73 378
1500 hours	250	167	59 525
Time average	415	276	74 705
Flow proportional	365	243	70 107

sample is taken after a preset number of increments of flow — say, 20 of 1 litre — have been measured.

The equipment will consequently sample more or less often as flow rate rises and falls. Some designs take a sample at regular intervals, and, triggered by flow rate, vary the sample volume taken. Most produce a composite sample of 1−5 l on a 24-hour rotating basis. Accurate loading calculations, percentage removals and mass balances are made possible and good quality data for future design becomes available.

The composite sample obtained is also often used by the regulatory body for compliance testing and charging purposes. They attend randomly to collect it and the factory is required to refrigerate and store the sample on a daily basis. As the equipment is measuring flow too, a useful control tool is now available to the factory manager to check water consumption, effluent generation and quality against production activity and time. A few flow-proportional samplers keep each sample taken separately, permitting individual examination by on-site staff. This may be useful for detailed surveys but poses an onerous analytical load.

In summary, investment in such equipment often proves a revelation as to effluent composition and provides a good opportunity to identify the source of the strongest, most polluting and therefore most expensive discharges.

Warren Jones Engineering supplies a wide range of sampling equipment, both fixed and portable, flow- and time-proportional. A typical installation is seen in Figure 7.1.

8.3.3 Time average

A series of equal volume spot samples are taken at constant intervals ranging from 5 minutes to 24 hours and either bulked into one bottle to form a composite or kept separate for individual examination. Where flow data are available, the samples can then be bulked-up manually on a simple proportional basis to obtain a rough flow-weighted sample.

The composite sample takes no direct account of changes in flow and for an effluent varying widely in strength will produce results that might seriously under- or overestimate the true load leaving the plant, particularly if the sampling frequency is low. Most battery-powered samplers take time-averaged samples but are capable of taking samples every 10−15 minutes. The large number of discrete samples comprising the composite therefore compensates for this area of potential inaccuracy.

This type of sampling can only be used with confidence where the effluent strength remains fairly constant irrespective of flow rate.

8.3.4 Manual spot sampling

Manual sampling must take place over a sufficient time period to capture the

Figure 8.1 Flow-proportional sampler/recorder.

peaks and troughs of plant operation as reflected by the liquid effluent. At the simplest level, spot samples can be taken with a can or bucket from a channel or well at random times day and night. A large number must be taken to build up a representative picture, as outlined in Section 7.7. The important factor is to randomise the sampling time as much as is practical and to use the opportunity to note activities, operational difficulties, etc, on-site at the time. Historical data collected in this manner should be viewed with caution; the individual circumstances of the site or plant may dictate sampling for 1–2 years before a reasonably representative picture emerges.

8.4 Sewage works sampling

Sewage works sampling is undertaken for three primary purposes:

(1) To satisfy legal requirements that the effluent complies with Consent limits, be it 95% or 100% of the time.

(2) To provide information to assist in the routine operation of the works.
(3) To build up a collection of historical data that will identify trends over long periods and so provide the basis of calculations for extensions or capital rebuilding programmes and mathematical modelling.

Whilst (2), above, is immediately useful in achieving the aims of (1), (3) is vital, if only to avoid at some future date, an ill-prepared and even unnecessary scheme or works based on poor quality or little data. It is clearly essential that a comprehensive survey of adequate duration is therefore carried out on any treatment plant for which major capital works are proposed, and where historical data is lacking in quality or quantity. The remarks about manual sampling above are particularly relevant here.

In the lack of in-house resources, this is an area of preparatory investigation that can usefully be offered to outside consultancies, being every bit as important as further planning, choosing a treatment option or a presentation at a public enquiry. More than one grandiose scheme has been based on too few samples, inaccurate analysis and incorrect interpretation.

It is unusual to translate the above-mentioned three purposes into separate sampling schemes, as the resources in many major water companies are often stretched to operate just one programme, although sampling for legal compliance is often carried out to a specific and quite separate programme from operational monitoring, and in the UK by a separate regulatory body, the National Rivers Authority (NRA).

Because of the degree of mixing, flow balancing and general buffering that will occur in all but the smallest sewage works and the capacity of the sewerage system to provide dilution, major and sudden changes of most parameters are unusual. Sampling programmes can therefore afford the luxury of a lower sampling frequency than in many industrial situations and time weighted sampling will prove accurate enough for operational purposes. The exceptions are freak weather conditions flushing out the sewerage system, or a shock load as a result of a spillage. The latter event dictates rapid pinpointing of the source, be it a road accident where a lot of fuel or chemicals may have been spilt or a problem at a factory or industrial estate. Either incident is able to considerably and detrimentally affect works operation and a lot of sampling is usually needed immediately so that decisions about appropriate action can be effectively taken.

At sewage treatment sites manned continuously, sampling round the clock can be carried out manually every two hours from three or four locations on site, and will provide reliable data for most operations while slowly adding to a data base. Samples can then be individually examined for odd colours, odours or precipitates. This is a particularly useful exercise for samples of crude sewage.

The reason for a sewage works rapidly going 'off' can be contained in the one 500 ml sample smelling strongly of cyanide or chlorinated solvent or coloured bright green by 500 mg/l of chromium sulphate. Alternatively, the individual samples can be bulked in equal volumes to produce time-weighted averages. If flow values are available, varying volumes taken of each spot sample in relation to flow will produce a flow-weighted composite, which can be used to determine plant loadings and percentage removals. If even-numbered hours are chosen one day and odd hours the next, the chances of missing the effects of recirculation, backflushing, desludging or some pumping activity will be reduced. At night, when flows are low and crude sewage weaker, such operational events can produce marked temporary shifts in effluent quality.

Most sewage works treating mainly domestic sewage exhibit a diurnal variation, the highest flows and concentrations entering the works during the morning, corresponding to the highest effluent flow rate and often concentrations closest to the Consent limits in the late afternoon, assuming 7–9 hours treatment and retention time in the plant. 9.30 AM is thus not the best time for repeated spot sampling of a works effluent unless you want to delude yourself as to actual works performance, as the inspectors of the regulatory bodies well know.

If carried out conscientiously, a tour of inspection of the site to take samples manually also allows observation and response to unusual events, breakdown or malfunction, and the opportunity to take further samples for analysis.

In the author's experience, and bearing in mind the size of many sites, a manual sampling frequency of less than two hours will prove onerous and will not be properly followed either by the works operators or on-site scientists. The sample bottles may well all come back to the laboratory full, but all containing the same sample taken just before heavy rain set in! But this will at least allow the astute chemist to conduct an informal check on analytical reproducibility and within-batch variation.

One of the few areas where manual spot sampling is essential concerns floating oil and grease where, for analytical purposes, deliberate concentration of the surface layer may be required. DO, chlorine, hydrogen sulphide (unless chemically fixed on site) and samples for parameters like sludge volume indices where gross denitrification may rapidly set in are other examples where immediacy of analysis dictates spot, manual sampling and nearby analysis.

Many sewage works sites are being progressively de-manned and samplers are being installed, emptied periodically by scientific or visiting maintenance staff. Most fixed samplers provide a bulked flow-weighted sample.

A recent development is to take individual samples – one per hour – and to continuously update after 24 have been taken by discarding the oldest. The sampler therefore always contains 24 samples covering the last 24 hours, providing individual samples for a routine monitoring programme and 'incident monitoring'. A further refinement allows sampler control by telemetry; when an incident is recorded, the sampler is switched to sample more often and may be switched off remotely to preserve the last 24 samples taken. This facility is equally relevant to the industrial scene and has been developed by Warren Jones Engineering Ltd.

It is important that samplers on site are regularly attended for testing, maintenance and sample collection. Refrigeration is highly desirable in warm or tropical climates as even high-quality effluent will significantly deteriorate in less than six hours and parameters such as BOD, NH_3N and NO_3N change enough to make operational decisions difficult. Conversely heating is necessary in cold environments to prevent uptake tubes and compressors freezing. Warren Jones Engineering Ltd and ISCO Environmental Division, USA, supply a wide range of samplers, composite and discrete, for many sampling environments.

Automatic sampling within sewers or manholes requires intrinsically safe (IS) equipment, as although not classified as Zone 1, it is informally acknowledged that sewers are hazardous areas and that equipment should either be IS or flameproof. Montec International Ltd supplies a range of portable samplers to this specification.

On a sewage works, the usual sampling locations are crude sewage inlet, primary sedimentation tank effluent, biological oxidation stage effluent and final effluent. Sludge samples during desludging or pumping should also be taken. Operational needs may dictate closer monitoring of particular areas: on large STWs, divided into sections, 12 or more sampling sites may be appropriate.

Smaller sites are better suited to automatic sampling using portable equipment; if resources dictate manual sampling, take crude sewage and final effluent samples as randomly as possible during the day and week and do not draw too many conclusions from less than 6 months' worth of data.

8.5 Sampling industrial sites

Industrial sites often contain a number of discharge points. If resources only run to manual sampling, each should be monitored on a random basis. While the 'I'm just passing so I'll take a sample' routine may sound random, it frequently is not; random number tables or even throwing darts at a board to decide is more promising. Alternatively, use the methods outlined in Section 7.7. Regular manual sampling can be carried out on a rotating basis providing the number of sampling points, time of day or week days are out of phase with each other. Avoid scenarios which include, 'It's 3.30 PM on Tuesday afternoon so it must be the sampling point in the basement!'

In view of the considerable variation in effluent volume and quality that occurs at many factory sites, however, manual sampling is not likely to provide a representative picture unless it is intensive and conducted over a very long time; automatic flow proportional sampling will provide a far more accurate picture.

Investment in automatic samplers often provides unexpected gains. Apart from the likelihood that a more accurate picture of the effluent will prove cheaper (see Table 7.1), the regulatory body will be able to collect a representative sample and see commitment to this and pollution control generally in the purchased hardware. They will be less likely to threaten legal action for accidental spillages, and the factory manager will have a first line of defence if something goes wrong in the factory and has the evidence in the bottles. Information wil! be available from the samples and flows to modify and tighten up on inefficient water and product usage and related work practices.

As for sewage works sites (see Section 7.4), fixed samplers on factory sites may require refrigeration facilities to avoid sample degeneration. (Warren Jones Engineering Ltd and ISCO Environmental Division supply appropriate samplers.) Samplers destined for enclosed areas or sampling effluents containing petrochemicals, hydrocarbons or traces of solvent should be IS. (Montec International supplies appropriate portable samplers.)

8.6 Sampling equipment and practical techniques

8.6.1 Manual

The hardware of manual sampling comprises a stainless steel can with handle fixed to a sufficient length of stout string or rope, preferably inert polypropylene or nylon. This will suffice in most cases. A can fixed to a rigid pole of wood or metal may be preferable with sludges or fast-flowing channels and easier to control. Plastic jugs or beakers can be used too; they tend to age harden and become scratched and are not suitable for trace analysis for organics, solvents or some metals because plasticisers and stabilisers slowly leach out. The fragility of a glass jar for capturing a sample precludes its use except with extreme care, although it is well suited as a sample container with a few exceptions mentioned below.

Whatever equipment is used, it must not contaminate the sample and must be flushed out with the liquid to be sampled before a sample is actually taken. This is particularly important when carrying out a sampling 'round' to avoid carryover.

Always sample the body of water or sludge as representatively as possible. In tanks, this may involve sampling at varying depths or across the surface. In channels, take as much of the flowing liquid without scraping the channel walls or any of the brickwork surrounds in a manhole. Flush out the sample bottle, then fill it completely and cap to exclude air that will cause oxidation and in time change the values of some of the determinands. To avoid further degeneration, refrigerate at 4°C or analyse the contents as soon as possible.

For the same reasons that applied to the sampling can, do not use plastic bottles for trace analysis involving solvents and organics and some metals (zinc and cadmium). Glass will be satisfactory for virtually all analysis except silica; use borosilicate glass if sodium analysis is proposed.

These simple aspects may sound very obvious, but expensive, highly accurate analysis is completely wasted on unrepresentative, contaminated and partially degraded samples and the resulting data will be of little use for control or design and may be expensively misleading.

8.6.2 Automatic

Automatic samplers certainly give a better picture of total performance than manual sampling confined to the working day, because the latter inevitably leads to an overestimate of BOD and SS loads to many treatment plant because incoming flow is usually weaker at night.

The permutations and programmes offered by even the simplest battery-operated samplers are many. Automatic samplers frequently have a delayed start feature so that specific operational activities are monitored; narrow time windows can thus be sampled at antisocial times or when site access is denied. Some equipment can be set such that sampling can be initiated only when a certain flow rate is exceeded, or only when there is a flow over a device like a weir. The latter is often used for storm water quality monitoring. Overlaying sampling has already been mentioned; the sampler operates continuously and discards and refills individual bottles. Normally, the previous 24 hours' worth are retained, a useful facility for event- or spillage monitoring.

Triggering sampling based on the pH falling outside a certain range or the presence of high solids or cyanide gas is unusual but feasible with automatic samplers capable of accepting an amplified signal from electrodes or sensors. Applications include monitoring trade effluent discharges to a pumping station, rising sludge blankets or providing evidence and extent of Consent exceedance.

Portable samplers operate from a 6 or 12 V battery of sufficient capacity to allow at least a week of continuous operation but ambient temperature and battery age will dictate exactly how long. Sealed lead acid batteries are commonly used; they must be charged up promptly at the correct rate and when not in use disconnected from the sampler. In the lack of a manufacturer's replacement, a motorcycle battery is an appropriate substitute or the sampler can be powered by a car battery for longer periods or via the charger or separate 12 V transformer/rectifier power supply.

Although fixed samplers are substantially housed in metal or grp cubicles and permanently wired and plumbed in, the portable type is designed to be carried a short distance, is frequently cylindrical in shape with a large carrying handle and weighs less than 20 kg. Cases are plastic and sample uptake tubing or external power supplies connect through external lockable sockets. Many designs are very compact and can be easily lowered into a standard size manhole opening.

In equipment designed to take a number of discrete samples, bottles are arranged on a rotating turntable with a fixed arm, though many recent designs use a robotic arm that moves to each bottle in turn. The bottles vary in volume from 250 to 1000 ml. The turntable is often attached to the sampler base as a separate item and can be quickly swapped for a composite sample bottle.

Most portable samplers take time-weighted samples and typically contain 24 bottles; with a little skill in setting up to avoid overfilling, four samples can be taken in each bottle, 96 in 24 hours. This is a very acceptable number for 'blitz' survey work. They can then be analysed separately, or manually mixed together to produce a time-weighted composite. The former allows individual examination – very useful at a sewage works that suffers spurious and toxic discharges – and for checking the effluent for spot compliance.

In cases where the liquid being sampled does not widely alter in composition and no flow figures are available, the composite provides a reasonable guide to average sewage or effluent quality and plant loadings and is often sufficiently reliable to form the basis for future design.

Some battery-operated portable samplers can be operated flow proportionally through a 4–20 mA signal input socket hooked up to a flow meter. Figure 7.2 shows a portable battery-operated sampler with 24 fixed 500 ml bottles and robotic arm. This will operate for about 10 days non-stop on a charge when set to take 48 samples/day.

Whether flow- or time-weighted sampling is chosen, it is essential that

Figure 8.2 Portable sampler on site.

neither the individual sample bottles nor the composite bottles overfills, otherwise concentration of solids will occur and the sample will not be representative. On portable equipment with a bottler unit where several samples are being collected in each bottle, mentioned above, some juggling with the sample volume collected each time the sampler operates is necessary. This can be difficult if the delivery head the uptake pump is operating against varies, as it often does in a well or sump. If in doubt, set the volume delivered low, and bulk several sample bottles together. Discard full bottles or regard their solids content as suspect.

For routine monitoring, setting the sampler to take samples every 15 or 30 minutes, bulking four into one bottle and so producing either 24 or 12 actual samples from 96 or 48 discrete samples taken covering a complete 24-hour period will produce good quality data on most sites. The samples can be individually examined and then further bulked manually on a flow-weighted basis if necessary. This sampling programme is a useful standard for any unfamiliar site or as the basis of an initial survey, as it is sufficiently comprehensive to enable interesting features to be spotted early on, and so allow changes that can concentrate on certain times or operational activities.

Remember that the sampler must be emptied regularly to avoid sample degeneration. In hot climates, this may require frequent attendance while in cool or wintry conditions, every 24 hours should suffice. At the same time, check for correct operation, pipe blockages, freezing up and displacement of uptake pumps or tubing. Ensure that pipes and bottles are kept clean to avoid carryover, particularly where the sampler is being moved from site to site. The action of the sampler, particularly during backflushing or purging, should not be allowed to disturb solids on channel walls or in shallow tanks and cause the subsequent sample to be high in solids (and probably other determinands).

Most automatic samplers will not take samples at a frequency of less than five minutes, by the time flushing out, collection and movement of the turntable or repositioning of the mechanics has taken place. It is unlikely that a pollution 'slug' of any significance would be missed at this sampling frequency or leave no trace in adjacent samples.

Limitations of automatic samplers include the number of bottles the sampler can hold, the size of each sample bottle and whether the resources of the laboratory are such as to allow prompt analysis of the samples as it is always tempting to have a sampling 'blitz' without realising the analytical effort involved. Most sample bottles supplied with automatic samplers are 500 ml or less, deliberately so in the battery-operated portable type in order to avoid their becoming too large to handle; 350 ml is the practical minimum volume for a reasonably comprehensive analysis suite of about 10 determinands to be undertaken.

As all contemporary bottles used in automatic equipment are made of some type of plastic, a further limitation relates to determinands. Glass should be used where analysis for pesticides, oil and grease, solvents and detergents is required. Volumes over 1 litre are usually needed to allow extraction, derivitisation or concentration before analysis. Samples for these determinands therefore have to be taken manually, unless a composite into a glass container is acceptable. Even then, the passage of the sample through the plastic uptake tubes of the sampler with the possibilities of carryover, adherence or chemical attack may not be acceptable.

Survey the site carefully before selecting sampling positions. Keep samplers away from heat, out of pathways and vehicle runs and place barricades round any partially-open manholes or pipes. Site the samplers at ground level in the open or a building where possible. Weight down uptake pipes to avoid sudden flushes displacing them but avoid blocking pipework; clean accumulated rag off frequently.

In view of the complexity of many factory sites and frequent initial uncertainty as to the precise drainage layout, a full site survey needs thorough and careful liaison with on-site staff. If simultaneous and continuous sampling at a number of points is proposed, portable automatic samplers are essential. Don't be afraid to alter and modify the survey as it proceeds and in the light of initial observations. In this respect, some early feedback of results from the laboratory is invaluable as they may indicate that sampling positions and the analytical parameters need changing. A copy of the drainage plan is essential to ensure that all discharges are captured or isolated where a mass balance or flow proportioning is required.

The human resources to operate a proper survey and the capital cost of samplers for this are often beyond the capabilities of many factory sites and best left to a specialist survey firm or consultancy. The latter should also be asked to provide a report outlining recommended improvements and modifications and covering areas of water and raw material usage and work practices.

8.7 Simple sampling statistics

It is important to define the purpose of sampling in order to establish when and how often to sample. Equally, it is necessary to build up a set of data before applying statistical techniques. Sampling objectives include the following:

(1) *Spot abnormal concentrations*. Continuous monitoring is required or at least the taking of a considerable number of samples at short intervals over a long period.
(2) *Identify peaks*. An intensive sampling 'blitz' will be necessary as in (1). The resources to handle the number of samples and data generated must be reckoned with.
(3) *Build up a picture of the process or monitor an established process*. For

this some knowledge of the frequency of variation within the treatment plant is essential. For example, an industrial plant would need monitoring over at least one production cycle or a number of shifts. If speed is not essential, a sewage works exhibiting normal diurnal variation could be sampled between one and seven times a week over a year or more for this purpose, taking one flow-weighted sample/hour, and bulking to produce a 24-hour composite. The sampling day of the week should be selected at random.

(4) *Estimate mean or average values*. While individual values are of interest in (1)–(3) above, composite samples will suffice for this objective.

(5) *Detect trends, change or deterioration in efficiency*. These are often long term and usually spotted in the results from a routine sampling programme, providing frequency of sampling is adequate. Changes may be seasonal as most plant work less well in winter and influent may vary widely (temporary increases in population at seaside resorts, production increases to cope with seasonal demands). Trends are often thought of as long term but in pollution incidents where the works receives a toxic shock load, intensive short-term sampling is needed.

(6) *Determine a fair industrial effluent charge*. The UK Water PLCs are progressively requiring flow proportional sampling facilities to be installed for this purpose. This provides a fair estimate of the average load leaving the factory site (normally measured as COD and SS) over the charging period (6–12 months) by virtue of the considerable number of individual samples comprising each composite.

As the equipment is automatic and not expensive, this provides the factory with a useful management tool to control water consumption and the content of discharges. Where previously samples have been taken apparently randomly and often only over the working day, the results of flow proportional sampling often come as a shock. Strength and costs have dramatically risen 'overnight'. In fact the skewed and unrepresentative nature of the previous arrangements are revealed.

(7) *Monitor for compliance*. In most situations effluent samples are expected to conform to standards 90% or 95% of the time with absolute 100% compliance values set at double (or at least substantially higher) than the 90%–95% ones. Sampling over the widest possible time window any day of the week, i.e. randomly, is the best defence against unpleasant revelations that the plant always fails after 15.00 hours on Monday. If you, the factory site manager, don't find this out, the staff from the regulatory body soon will!

In summary, if there is little or no diurnal or day-to-day variation, sampling may be randomised by time and carried out evenly over a year or more. Sampling for peak loads should be carried out when these are known to occur from the results of preliminary investigations. Sampling industrial effluent should encompass a production cycle or batch process and seasonal variations. Sampling to spot trends needs the elimination of diurnal and day-to-day variations and should be carried out on the same day each week if, for example, month-to-month trends are to be spotted.

To determine the days of sampling in a programme lasting one year, random number tables may be used, selecting a number between 365/n and 0, where n is the number of samples and exceeds 25. If n is less than 25, choose a number between 52/n and 0 to find which week number to sample in. Some computer programs will generate random numbers and a few calculators also have this facility. The dartboard has already been mentioned; do not let your star player have all the throws!

Formula 1 below indicates the day number during which sampling should take place.

$$A + 365/n, A + 365 \times 2/n, A + 365 \times 3/n \ldots A + 365 \times n/n \quad (1)$$

where:

n = the number of samples (larger than 25)
A = a random number in the interval 365/n and 0.

Formula 2 applies where n, the number of samples, is less than 25 and B is a random number in the interval 52/n and 0.

$$B + 52/n, B + 52 \times 2/n, B + 52 \times 3/n \ldots B + 52 \times n/n \quad (2)$$

A similar formula can be used for other periods — 1, 3, 6 months.

Having generated the sampling dates, check through the calendar to ensure that they do not include holidays, shutdown or nonoperational periods, and that inadvertently the days chosen don't all happen to be Wednesday or during a cleaning-down cycle.

Deciding the number of samples n to take requires preliminary sampling to provide the values for statistical techniques to be applied. As all effluents vary in quality either randomly or systematically, the values obtained such as mean, minimum and maximum and standard deviation are only estimates of the true values and usually different from them. Where the variations are random, the differences between the true and calculated values decrease as the sample numbers increase. As it is not practical to take vast numbers of samples, and assuming cyclic variations are either absent or small compared with random variations, the number of samples taken need only be large enough to meet acceptable uncertainty at the confidence level chosen.

The confidence level, e.g. 50%, 90% or 95%, is the probability that the true mean $\bar{x}$ of n results will be included within the calculated confidence interval L. The confidence interval L for the calculated mean value $\bar{x}$ of a particular determinand on the basis of n samples at the 95% confidence level means there are 95 chances in 100 that the confidence interval L will contain the true mean $\bar{x}$. Where a large number of samples is taken, the frequency with which L includes $\bar{x}$ will be close to 95%.

For a number of (sample) results n, taken at random, estimates of the true mean $\bar{x}$ and the standard deviation σ are the arithmetic mean $\bar{x}$ and s respectively as in the formula:

$$s = \frac{\sqrt{\sum_{i=1}^{n}(x_i - \bar{x})^2}}{n - 1} = \sqrt{\frac{1}{n-1}\left[\sum_{i=1}^{n} x_i^2 \frac{1}{n}\left(\sum_{i=1}^{n} x_i\right)^2\right]} \quad (3)$$

x_i represents the individual values

If n is large, σ and s are little different and the confidence interval of $\bar{x}$ from n results is $+/- K/n$ $\bar{x} \pm K/n$.

K has the values given in the table below, depending on the confidence level chosen.

Confidence level (%)	99	98	95	90	80	68	50
K	2.58	2.33	1.96	1.64	1.28	1.00	0.67

To estimate the mean $\bar{x}$ for a given confidence level L at the chosen value of K, the number of samples necessary is $(2K\sigma/L)^2$. In order to apply this formula σ should be accurately known.

Assuming normal distribution, i.e. about 95% of sample values lie within two standard deviations of the true value, the confidence interval L of the mean of n results can be calculated by rearranging the formula:

$$L = \frac{2K\sigma}{\sqrt{n}} \quad (4)$$

As an example, if the confidence interval was 10% of the mean, i.e. the true and calculated mean were within 10% of each other, the required confidence level 95% (there is a 95% chance they are within 10% of each other) and the standard deviation 20% of the mean, then substituting from Formula (4) above:

$$10 = \frac{2 \times 1.96 \times 20}{\sqrt{n}}$$

So $\sqrt{n} = 7.84$

$n = 61$

Nine samples/day are required if the period of interest is one week, two/day if one month and one or two/week if sampling over one year.

At the higher confidence level of 99% and 10% confidence interval, $n = 106$. If the sampling period is just one day and the day chosen is an average representation of any other day of the week in effluent terms, it is possible to set an automatic sampler to take 96 samples, one every 15 minutes, and obtain statistically an apparently accurate picture of water quality.

Conversely, at the sacrifice of confidence (50% level), only seven samples are needed and a minimal effort to obtain them. However, this example probably serves better than most to illustrate the low confidence that can be placed on results by taking a sample of the effluent only once every two months and averaging a year of results. It would be a far better use of such limited resources in most cases to have an occasional 'blitz' – e.g. one week once a year, always assuming the chosen week is a representative one and day-to-day variations are minimal. This assumption might apply on a small sewage works receiving domestic flows but is relatively uncommon in a manufacturing environment.

8.8 Flow measurement methods and equipment

Accurate flow measurement at a treatment plant or of the discharge to the foul sewer or watercourse is essential for the following reasons.

(1) Most regulatory bodies place upper discharge rates/hour and a 24-hour total flow to avoid flooding and surcharging of pipes and chambers, inadequate dilution of the effluent in a river or estuary, or treatment difficulty at the local works caused by sudden shock loads. It is often a condition of a Consent to Discharge that flow is measured continuously.
(2) Flow measurement clearly forms an essential part in arriving at costs/m^3 in a charging formula applied to industrial discharges.
(3) Recirculation rates, pumping and diversion weirs are often controlled automatically by feedback loops generated by flow recorders and loggers.
(4) Flow records will be necessary for any mass balance calculations through a plant and will greatly assist water consumption studies.

The equipment is often situated in an open channel of known dimensions and measures the liquid height variation of the flowing liquid. Older installations comprised a float which either directly operated a pen on a slowly rotating paper chart or by mechanical linkage operated a totaliser. Virtually all modern equipment measures height variation ultrasonically at a frequency of 60 kHz. The sensor head is carefully located over the liquid so that spurious reflections are not picked up from channel walls and is aligned horizontal to the liquid flow.

The recorder contains a number of programs to allow flow to be measured in a variety of environments. Typical options include flow over a rectangular or V-notch weir, in a flat or U-shaped channel upstream of a pre-shaped flume or across a river of (infinite) width. The recorder also stores the data – typically readings are taken every 10–60 seconds – and will download it directly upon interrogation into a laptop computer or logger in a number of configurations including tabulations and graphical display. Fixed-flow recorders often transmit flow data at intervals automatically by telemetry. The memory of most recorders will store at least 30 days' worth of data and is maintained by a separate back-up battery which also maintains a real-time clock. When the main operating 6–12 V battery is discharged, recording will cease but the data remains protected until the equipment is retrieved and recharged.

The modern generation of portable recorders is compact and may be carried easily. Figure 7.3 shows such an installation *in situ* supplied by Warren Jones Engineering Ltd. Fixed versions are little larger and can easily fit into cabinets originally designed for the earlier mechanical flow recorders.

All flow recorders must be carefully installed. Follow the manufacturer's instructions and pay particular attention to accurate measurement of distances and heights of structures and water levels, for which an accurate, rigid ruler at least 1 m long is essential. Often the most difficult parameter to measure is the liquid height flowing over a weir as it tends to fluctuate quite widely and rapidly and some perseverance may be required to input this critical value accurately.

Figure 8.3 Portable flow recorder on site.

The beam angle of most sensors is only about 6%, allowing installation in narrow channels. However, the sensor has to be located some distance above the channel or liquid (more than 200 mm) and at some areas of a plant it may prove impossible to measure flow unless a purpose-built weir or flume can be inserted.

This should be an essential requisite in any new treatment plant design for it is virtually impossible to measure flow accurately in the confines of many manholes and equally so in a closed pipe either half full of liquid, or surcharged but where vortexing occurs.

Fortunately, most wastewater treatment plant have at least one open channel containing a flume, or discharge weir for flow measurement. For difficult locations where neither are present, ISCO Environmental Division supply a variable gate flow meter comprising a metering insert and a pneumatically controlled gate which creates an upstream level artificially and measures the flow by a bubbler system. To prevent rag and silt build up, flushing and air purging are automated features of this equipment.

8.9 Effluent charging methods

Until recently, many discharges of industrial effluent, particularly those to estuaries or sewers that discharged untreated sewage to the sea, enjoyed very low fixed prices and no account was taken of either volume or strength. The Urban Wastewater Directive is changing this position in Europe as the regulatory bodies recover their operating costs, and worldwide, a more rigorous application of the 'polluter pays' philosophy means that firms still in this position are living on borrowed time and must anticipate a steep rise in future effluent costs.

In the UK and much of Europe and the USA, two factors are involved in charging for discharges of trade effluent made to the foul sewer and in some countries permutations of these apply to effluent discharges to watercourses. The two factors are:

(1) Flow measured in cubic metres or gallons.
(2) The strength of the effluent. Usually solids content and oxygen demand are the two parameters used to measure 'polluting strength'.

In the UK, Settleable Solids (SS Settled) and Chemical Oxygen Demand (COD) are measured in samples taken by the regulatory body and applied in a charging formula where they are compared against national average values for the same determinands measured in domestic crude sewage. The formula also contains fixed values for sludge disposal, pumping and reception, biological oxidation and primary treatment per cubic metre; these are reviewed annually. The formula applied by the UK Water PLCs is detailed below.

$$\text{Cost in pence/m}^3 = R + V + \frac{Ot}{Os} B + \frac{St}{Ss} S$$

Where:

R = the average sewerage, pumping and reception costs/m^3.
V = the average cost of primary treatment/m^3.
Ot = the COD of the trade effluent in mg/l after one hour settlement at pH 7.
Os = the COD of average strength settled crude sewage in mg/l.
B = the average cost of biological oxidation/m^3.
St = the settleable solids of the trade effluent in mg/l after one hour settlement at pH 7.
Ss = the settleable solids of average strength settled crude sewage in mg/l after one hour settlement.
S = the average cost for treatment and disposal of primary sludges per cubic metre of sewage.

Sections of this formula may not apply to some discharges and the values for the factors R, B, V and S will vary between different companies to reflect their individual operating costs. Ot and Os are the national average values mentioned above and vary slightly from year to year. Note that all the chemical analysis refers to one hour settled values for COD and Solids.

Bills are sent out every 6 or 12 months. Large dischargers may be invoiced at 1 or 3 month intervals.

While this formula has been in existence for some years, banded charges and a simpler formula are likely in the future to reduce administration costs. They will not necessarily prove fairer to the discharger whose present effluent charges hover about the transition point between bands. The formula also takes no account of toxic components in the discharge, e.g. metals, which are permitted at low levels, and it is likely these will carry some added cost weighting in the future to reflect treatment difficulties and restricted sludge disposal options incurred by the receiving body.

In principle, the formula is fair; if the trade effluent is ten times the strength of crude sewage (typically £0.5–£1.0/m^3 to fully treat at 1994 prices), the discharger will be charged 10 times this rate and it will be obvious that any measures that can be taken to reduce volume and strength of effluent will reap a financial bonus.

In order to cover its operating costs fairly, the receiving authority will want to take representative samples of effluent and for this reason flow-proportional composite sampling is becoming the norm rather than reliance on working hours spot samples which can give a highly biased picture in many manufacturing scenarios, usually to a factory's advantage.

Some firms have already discovered this to their cost, the effluent bill having risen by 5–10 times. In these circumstances, it really is time to carry out a properly conducted survey in order to review water consumption, work practices, waste production and effluent generating activities; necessary changes should then be implemented quickly.

In the UK, the Water Industry Act 1991 Part IV Section 121 allows the Water PLC to insist on the installation of flow-proportional sampling and flow measurement equipment at the discharger's expense.

8.10 On-line sensors for continuous monitoring

In many process industries much routine sampling and laboratory-based analysis can be dispensed with, where the environment enables reliable on-line continuous monitoring to be carried out using the range of sensors now available. For monitoring wastewater, sensors with a rugged industrial specification are available to measure DO, turbidity, suspended solids, pH, temperature, conductivity and chloride directly in the flow.

Ammonia (both as the ammonium ion NH_4+ and NH_3N liberated as the gas from an alkaline solution), phosphate, silica, fluoride, nitrate and colour are further determinands that can be measured remotely. For these the sensors are usually housed in a custom-built monitor along with reservoirs for reagents and standards, a sample being periodically pumped to the equipment from the effluent stream during an entirely automated cycle of flushing, calibration and measurement.

ABB Kent Taylor Ltd supplies a number of purpose-built monitors and *in-situ* sensors and electrodes worldwide; ELE International Ltd specialises in DO and pH electrodes of exceptional ruggedness that will withstand many months of continuous operation in the harsh environment of a sewage works aeration plant.

Successful on-line monitoring relies critically on the fulfilment of several requirements.

(1) Adequate maintenance and servicing of the electrode membrane and any filling solution, and periodic cleaning.
(2) Frequent, accurate calibration.
(3) An understanding of the characteristics of the particular electrode in use – interferences, susceptibility to drifting and detection limit and measuring range.

Again, the manufacturer's instructions should be followed carefully and specific guidance is not appropriate here, but there are several practical points to note. In systems where the sample is withdrawn to a separate cubicle for analysis by pump, tube blockages are common from a solids/rag build-up and wide bore tubing and a fast pumping rate are required. Strainers placed on the end of the sample uptake tube can be more trouble than merely dangling the tube in the channel and letting it move with the flow. This will often prevent rag accumulating, particularly where the tubing is smooth-walled and flexible.

For the same reason, the author has often found it better to let electrodes hang freely in the flow rather than rigidly fixing them, particularly on sewage works where rag is a constant problem in aeration plant or the preliminary screening facilities are inadequate. The electrode should be completely immersed all the time and depending on flow velocities, some weights may be necessary.

Maintenance and cleaning frequencies can only be determined by experience. For instance, DO electrodes may survive for 12 months before a new membrane and light cleaning of the silver anode with fine emery cloth is required to restore performance. In corrosive environments or those with high sulphides, physical attack of electrodes may be apparent within one week. SS electrodes have been known to drift out of range in 15 minutes where the cell path length is too short. It is therefore vital, after installation, that the electrode is regularly examined and relocated if necessary.

To avoid interference problems, a wide-ranging analysis of the effluent or flow to be monitored should be carried out first and determinand levels checked against the manufacturer's list of known interferences for the electrode. It may then be necessary to withdraw a sample for dilution before measurement, or if the interference is constant and the same concentration, to prepare a correction chart or graphs.

High levels of chloride are frequently problematic; surface active materials (detergents) can also markedly affect the membrane characteristics of selective ion electrodes and cause elevated readings.

As a general rule, an external calibration of the electrode should be performed by a competent person at least once a week. DO electrodes should be withdrawn, washed and swung in air to obtain the 100% saturation value, then immersed in 5% sodium sulphite solution to obtain zero. Most other types are calibrated against freshly made standards; often the manufacturer can supply these and the HMSO/SCA methods series, contains instructions for making up solutions of known concentration. This practice should be adopted for fixed analysers too; calibration standards will drift and degenerate in high temperatures and the reservoirs of these should only be sufficient for two weeks of automated analysis. All electrodes should respond sufficiently rapidly to achieve 90% of true value within one minute; failing this, they require cleaning and possibly a more general service.

Be careful not to install electrodes to control operational activities where the measurable concentration is close to the detection limit. Drifting, poor reproducibility and long response times may occur. Sample withdrawal followed by pre-concentration may be necessary, although exceptionally, most DO electrodes will work for long periods at 5%–15% saturation without ill effect.

Remember that on-line monitoring is not an 'install and forget' method –

the quality of the data generated depends on the measuring accuracy of the sensor(s) and there is no sense in having £1 million worth of plant controlled by electrodes that have not been calibrated, cleaned or even checked for their very existence for 6 months.

8.11 Analysis: initial considerations

Careful consideration must be given when choosing determinands to form an analytical suite. Apart from the obvious observation that hastily chosen analysis may not tell you what you want to know about, such a lack of consideration coupled with taking only a few samples may prove disastrous where plant modifications or a new facility are proposed.

Formal laboratory analysis is expensive and the on-site laboratory represents a considerable investment in staff and equipment, although in the industrial situation, the laboratory function is primarily one of quality assurance. Sampling and analysis of the effluent will often provide a welcome break from daily routines for the staff and allow an expansion of analytical skills. Most of the basic tests for water pollution can be performed with a modest outlay in glassware or electrodes.

The most important thing is to adopt good, consistent technique and use well-documented methods. Almost all wastewater analysis is carried out at the mg/l or µg/l level; hence methods are quite specific, well documented and not likely to be improved by experimental modification.

Over 50% of the cost of a survey or routine sampling programme is often accounted for when the analysis is contracted out. The quality of the chosen laboratory must be checked:

- Are the methods regularly calibrated and do they maintain a comprehensive system of analytical quality control (AQC) and good laboratory practice (GLP)?
- What are the sample storage and data handling facilities and when will you get the results?
- Are the staff appropriately qualified and are the charges reasonable?

Many commercial laboratories and those belonging to former public bodies have joined a national accreditation scheme in which they are annually inspected by a competent, independent body in many areas of activity, particularly AQC and data handling. In the UK, the National Measurement Accreditation Service serves this purpose.

Very little environmental analysis is carried out 'blind'. Usually, it is possible to have a good idea of likely components of wastewater providing the source is known, which is why the first line of action in a pollution incident is accurate identification of the source, or in a factory the contributing flows.

If there really is no preliminary knowledge, a general sweep of determinands that measure oxygen demand, solids levels, pH and, depending on visual appearance, colour and odour, metals or solvents will provide useful pointers. If the resources and finances allow, gas chromatography/mass spectrometry (GC/MS) will provide the answer as to the composition of a cocktail of organic compounds. In many routine situations, the approximate concentration of each determinand will also be known amd the method used can be calibrated over a suitable range.

What level of accuracy do you expect to be achieved during analysis? There is little justification spending resources on good quality sampling if the laboratory can only produce answers ±50% of the true value, reproducibility is poor, the method is not validated by good AQC or calibrated over a range outside that of normal determinand concentrations.

Equally, answers to three decimal places are unjustified in most wastewater operational situations where no extra control can be achieved by such a level of accuracy and the very nature of a constantly changing wastewater negates such apparent 'accuracy'. Few methods will reliably reproduce answers where the second decimal place has any meaning. Statistics as applied to analysis are well documented in the HMSO/SCA methods and form the basis of AQC. The reader is referred to these, and for those with a good understanding of maths, there are a considerable number of books on the market on this subject.

Some tests will not be appropriate and may give misleading answers. One specific determinand, the 5 day BOD, primarily measures oxygen consumption through biological activity. When applied to industrial effluents containing inhibitory substances, it can give misleading, nonreproducible, low or even nil values, and is consequently rarely specified in Discharge Consents unless the wastewater or treatment plant relates to the food industry. The COD or TOC test is a better choice to assess potential oxygen demand and both are much quicker to perform.

Be aware of interferences; it may be necessary to employ the method of standard additions, where different measured amounts of a standard are added to aliquots of the sample and a graph constructed to obtain the sample value by extrapolation.

Laboratory electrodes are just as prone to giving erratic or seriously modified values as the on-line variety mentioned in Section 7.10. A possible solution to this problem, which can be applied more easily in the laboratory than on-line, is to dilute the sample to the point where the effect of the interference is either negligible or constant, and apply a correction if necessary. If the interfering substance is not constant in concentration, it will need to be measured too and a series of correction look-up tables or graphs prepared.

All wastewater and sludge samples must be analysed as soon as possible after collection. Degeneration caused by temperature and light will seriously alter determinand values within a few hours and transportation to the laboratory must be efficiently organised to minimise delay. The BOD test and the NH_3N/NO_3N ratio are particularly sensitive.

Always choose a laboratory local to the sampling site if contracting out analysis. For a few determinands, it is possible to 'fix' the determinand value in a bottle at the time of collection. DO preserved as manganous/manganic hydroxides is an example where the DO may change rapidly in 30 minutes if the sample contains bacteria with a high respiration rate. Once at the laboratory, refrigeration at 1°C–4°C will arrest further degeneration until analysis can start; on-site cooling at the point of sampling is highly desirable in tropical conditions.

8.11.1 Choosing analytical suites

Routine sampling or surveys are rarely aimless and purpose often conveniently defines the analysis required.

(1) If to test compliance with Consent limits, clearly those in the Consent will be chosen – BOD, COD, SS, pH range, toxic metals. It is a sensible precaution to test randomly for prohibited substances, e.g. solvents and fuels where these are in use on site.

Trade effluent Consents often contain specific parameters, e.g. carbohydrate, formaldehyde, phenols, chloride, iron, grease and oil which the Regulatory body will know are present on-site in quantities that, if discharged, could cause treatment difficulties at the receiving works. Similarly, treated effluents discharging with low dilution to a watercourse may have DO, nitrate and phosphate limits imposed (as well as the conventional limits on BOD, SS and NH_3N) in order to prevent ecological imbalance.

(2) Monitoring plant performance to build up a data bank will, in time, provide the operating range of the plant in terms of oxidation or specific parameter removal and spot trends. Depending on sampling frequency, abnormal concentrations will also be identified.

As removal efficiency is often of primary interest, analysis will be concentrated on reductions of BOD, COD, SS and NH_3N and conversely the increase in NO_3N and DO. pH is a relatively cheap and accurate addition, while phosphate, nitrite and sulphate or sulphide may also be appropriate. Metal removal rates during sedimentation and fat removal by flotation are further examples in industrial plants.

Primary focus should be placed on the influent and effluent, the former usually exhibiting the widest range of concentrations. BOD, COD and the NH_3N/NO_3N ratio are good choices to monitor the intermediate stages of oxidation. If the analysis results exceed 90% accuracy and the sampling regime is statistically sound, some satisfying mass balances through the plant will be possible.

(3) Surveys designed to provide data for new plant or modifications to present facilities will need the range described in (2) above, but if no serious monitoring has been carried out before, a much wider suite that includes

analysis for the raw materials used in the factory will be useful for influent samples. This is particularly so if a new treatment system is likely to receive toxic solvents, organic cocktails, metals or sporadic oil discharges. A review of manufacturing activities will therefore be necessary and some surprising discoveries of wasteful activities may be made. At the same time, it would be prudent to include parameters that the regulatory body may be seeking to impose in several years' time, or are going to include in the charging formula for the discharge.

Sludge analysis should not be overlooked; DM%, and others relevant to treatment and disposal such as N, P, calorific value and O & V% are important. If anaerobic digestion is proposed, a scan for toxic metals or solvents that might hinder digestion is essential and routine samples should be taken during normal operation to check that the next batch of feed sludge is not going to cause problems.

8.11.2 Common determinands

A brief description of the more common determinands measured in wastewaters and their relevance is given below for the benefit of non-chemists. The HMSO/SCA methods provide a definitive description, analytical method, standards and performance of these and many more and the reader is strongly recommended to refer to these or to similar standard and reported methods in their own country.

Biological Oxygen Demand (BOD)

This measures the oxygen demand of both the organic matter and organisms living in the sample and has been in use since the 1920s as a basic measure of the polluting effect a discharge has on a receiving watercourse.

Most sewage works consents worldwide contain this parameter. The test takes five days, the oxygen content at the start and after five days incubation in the dark at 20°C being measured by DO electrode or titration. 250 ml glass bottles are normally used. The 20°C temperature standard represents an average value for slow moving streams in temperate climates and is easily duplicated in the laboratory incubator.

Most samples will require dilution with air-saturated water containing nutrients to promote microbial activity if the oxygen requirements of the biomass is to be satisfied without 'stripping' (to nil oxygen); and some may need 'seeding' with a good quality effluent if microbial numbers are low or absent as industrial effluents often are.

Dilutions from 1:5 to 1:10 for good quality treated effluents and 1:100 for crude sewage are typical; the discharge from a brewery or food process may require a dilution of 1000+. Treated effluent will have a BOD of 5–50 mg/l, crude sewage 250–600 mg/l and organic industrial effluents may exceed 50 000 mg/l.

Be aware that the presence of toxic materials may give very low or nil BOD values by inhibiting microbial activity; the unsuitability of this test for some industrial effluents has been mentioned in earlier sections and must be an individual decision for each discharge.

Perhaps of all wastewater analysis, the BOD is most demanding of consistent technique which must be maintained to achieve acceptable accuracy; the author has witnessed much individual variation between fellow chemists performing this test! In particular, the dilution water can often be a source of high blanks. It will readily adsorb vapours in the laboratory during aeration, can be contaminated by oil from the compressed air source and may supersaturate. These are all points to check if blanks greater than 0.5 mg/l are measured.

The detection limit of the BOD test is about 2 mg/l. Values below 5 mg/l are not likely to exhibit better than ±20% reproducibility. Despite the disadvantage of having to wait 5 days for the answer, the BOD remains an unousted industry standard for measuring the general operating efficiency of a sewage treatment plant and seems set to retain that status.

Chemical Oxygen Demand (COD)

The sample, diluted as appropriate, is boiled with a mixture of sulphuric acid and potassium chromate. A catalyst and chloride suppressant is often added too. Small volumes are involved (less than 30 ml) but safety precautions are necessary. The test, including an end titration with ferrous ammonium sulphate and an indicator (ferroin), is complete within three hours includng cooling down.

Although refluxing with chromic acid mixture at 360°C hardly emulates average conditions in a wastewater plant or the local river, the COD test provides a rapid and useful indication of the likely total oxygen requirement of the sample without relying on the sometimes fickle behaviour of microbes in dynamic second and third order reactions (the BOD test). A few refractory compounds remain untouched by this aggressive oxidation and the effluent from incineration processes may give slighty depressed values. Typically, the COD is one and a half to three times the BOD value, and this ratio can give a useful feel of the 'treatability' by conventional means of a wastewater; the higher the ratio, the more resistant the effluent will be to biological oxidation.

The COD produces consistent results and a high level of accuracy. The practical detection limit is about 5 mg/l. Good accuracy and reproducibility should be achieved if the reagents are fresh and stored at 4°C. The generalised reaction proceeds thus

$$\text{Organic matter } (C_A H_B O_C) + Cr_2 O_7^{2-} + H^+ \xrightarrow[\text{heat}]{\text{catalyst}} Cr^{3+} + CO_2 + H_2O$$

It is widely used in charging formulae for trade effluents (see Section 7.9).

Suspended solids (SS)

The sample is sucked by vacuum through a preweighed circle of filter paper. Volumes of 25–500 ml are used, the suspended solids level and therefore the filtration time being deciding factors. After drying at 105°C for 1–2 hours, the paper is weighed again, the difference being related to the sample volume passed through.

Good quality effluents will have an SS of less than 15 mg/l, while crude sewage is 250–400 mg/l. Often the solids/BOD ratio is about 2:1; some industrial food wastes have very high and similar SS and BOD values. They will be readily treated biologically but will not benefit from initial settlement because much of the BOD is in the soluble phase.

The standard filter circle is a 7 cm circle of GF/C paper, a glass-fibre material with particle retention down to 1.5 μm. The majority of solids particles in some sewage works effluents are less than 10 μm, while the average microstrainer or drum filter only retains solids greater than the installed 22–35 μm mesh. These differences are worthy of note in cases where a plant fails Consent on SS alone and a fine mesh screen is proposed to solve the problem. Sludge dewatering by vacuum is a major source of particles less than 10 μm or 'fines' but they respond to coagulation by polyelectrolytes and cosettlement with raw sludge.

GF/C papers often have variable quantities of loose fibres trapped between them, and flicking through a batch followed by washing often works wonders for the laboratory AQC charts by reducing spurious 'fliers' in otherwise steady values. A filler of zinc oxide is commonly used to bind the fibres, so metal analysis of the filtrate is not advised and conventional filter paper should be used.

This is an accurate and quick test that provides much useful information about the effectiveness of settlement tanks, solids production and mass balances within a treatment plant and loads to a watercourse. In common with COD, it appears widely in trade effluent charging formulae and limits of 300–500 mg/l are often applied to avoid sewer blockages and excessive sludge volumes at the receiving works. The detection limit is about 1 mg/l and reproducibility very good if the balances used for weighing are regularly checked and kept clean.

Ammonia (NH_3)

Ammonia arises in wastewaters from the breakdown of amines, proteins and nitrogen compounds and the hydrolysis of urea. It is always present in sewage within a typical range of 20–50 mg/l, the 'stronger' the sewage the higher the ammonia. Results are usually expressed as ammonia nitrogen (NH_3N). Farm wastes and sludge dewatering liquors often contain 100–200 mg/l NH_3N and such levels often require pretreatment or more oxidation capacity to be installed.

Ammonia is toxic to fish as the unionised ammonia fraction (a parameter that can be calculated from a nomograph containing the two other factors, pH and temperature) and hence it frequently appears in effluent discharge Consent standards, values of 2–10 mg/l being common in Europe.

Trade effluent Consents for organic discharges often contain an ammonia standard to prevent odour problems at pumping stations and overloading the oxidation capacity of the receiving works.

Most removal occurs in the activated sludge or biofilter section of treatment, where NH_3N is oxidised to nitrite and nitrate by nitrifying bacteria preceded by carbonaceous oxidation. Thus, an overloaded or poorly-maintained plant will only achieve partial nitrification or only carbonaceous removal and the ammonia test provides a fundamental assessment of the efficiency of oxidation: 60%–90% removals are common in conventional sewage treatment, filter plant achieving noticeably less in cold winter weather while oxidation ditches often exceed 95%. Thus, a good quality effluent contains less than 5 mg/l NH_3N, less than 1 mg/l being normal for oxidation ditches and even filters under favourable, warm conditions.

Unless the plant also denitrifies, the ammonia will have been converted to nitrate, and thus the sum of the ammonia and nitrate in the final effluent should roughly equal the ammonia at the crude inlet. There is rarely any nitrate in sewage or wastewaters of comparable strength unless as a gross input from, say, farming (and even here, nitrate is rapidly reduced to nitrogen). Roughing filters or purpose-built ammonia stripping towers are usually designed to remove 50% of the NH_3N from wastewaters with a high soluble BOD content and liable to rapid degeneration.

Raw sludges treated with lime will release NH_3N to atmosphere readily; conversely, wastewaters low in nitrogen frequently contain no measurable ammonia, dairy industry effluent being one example where nitrogen as urea often has to be added to the inflow to the oxidation unit to foster satisfactory biological nitrification and carbonaceous breakdown.

The colorimetric techniques still used in auto-analysers to measure ammonia are now complemented by ion-selective electrodes that can either measure the ammonium ion NH_4^+ or ammonia expressed as nitrogen (NH_3N) stripped from solution as the gas, for which purpose the sample has caustic soda added to it first and the pH raised to at least 12.

While the ammonium ion can therefore be measured *in situ* with portable equipment (ELE International is a supplier of this), on-site fixed analysers (ABB Kent Taylor Ltd is a supplier of these) usually measure NH_3N as the gas and withdraw a sample from the flow for analysis after reagent addition. Such equipment requires regular maintenance, calibration and cleaning over and above any automatic facilities, as described in Section 7.10.

The gas sensing electrode was first introduced in the early 1970s and has been engineered to an industrial robustness. In a sewage works environment where fat, grease and rag are constant problems for any on-line sensor, the membrane of the ammonia electrode is the most vulnerable area and needs regular examination if results are to be deemed reliable.

Detection limits in the laboratory environment are 0.1 mg/l or better – more than adequate for reporting wastewater analysis. On-site measurement produces reasonably fruitful answers down to 0.5 mg/l but reproducibility may vary; this should be borne in mind if a very tight ammonia standard is imposed on the effluent and equipment regularly serviced.

In common with many wastewater analytical parameters, the ammonia value of a sample starts to change from the moment it is taken, as a result of organic degeneration, pH and temperature changes and the action of light. If nearly immediate analysis is impossible, samples must be refrigerated to 4°C or less and kept in the dark. This consideration applies equally to the next parameter, nitrate.

Nitrate (NO_3)

This parameter forms the 'other half' of the ammonia analysis, the ratio between the two indicating the degree of oxidation being achieved during treatment. Similarly, results are usually expressed as nitrate nitrogen NO_3N. Again, sensitive laboratory colorimetric techniques will detect 0.05 mg/l, and on-line measurement by electrode achieves 0.2 mg/l. Reproducibility is often excellent.

Some electrodes are, however, severely affected by interferences from chloride and surface active materials (detergents) which are frequently in wastewaters, and the manufacturer's performance literature must be studied carefully and a wide ranging analysis conducted first if meaningful on-line measurements are to be made. Many manufacturers, including ABB Kent Taylor Ltd, supply the electrode with a supply of sensing capsules which are merely replaced as necessary and the task of positioning membranes and replenishing filling solutions is dispensed with. Capsules may last up to 6 months.

A well-nitrified sewage works effluent typically contains 25–40 mg/l NO_3N, while an oxidised and denitrified effluent less than 5 mg/l. Poor quality effluent or that from a plant only performing carbonaceous oxidation also contains little nitrate. Although a contributing factor to eutrophication in rivers and streams, nitrate standards are rarely applied to treatment plant at present; ultimately, a nitrogen load limit may be imposed in sensitive areas for which both NH_3N and NO_3N measurements will be necessary.

pH

This is perhaps one of the easiest parameters to measure by using either pH papers that provide a colour change over chosen ranges of pH, or electrode. Many industrial versions of the latter are available and deliver reliable results for long periods without attention.

Both ABB Kent-Taylor Ltd and Mettler Toledo Ltd are suppliers of in-line pH electrodes and a number of manufacturers supply portable pH electrodes coupled to an LCD display. Whatman Labsales Ltd sells a typical range of styles with a resolution of 0.01 pH and an accuracy of 0.02 pH. Many factors influence the pH of wastewaters which are normally a complex mixture of cations and anions, but successful biological treatment is confined to a narrow range — typically pH 6.5—8.5. Because of this, pH monitoring/alarm facilities are desirable at the influent to plants likely to receive acidic/alkaline inputs and where the buffering or dilution capacity is limited; pH correction should be engineered as reliably as possible. High/low pH values often indicate that other undesirable components like toxic metals or solvents are present; unlike simple pH these are difficult to measure in-line. Measuring this parameter often proves a good investment against sudden plant failure.

8.12 Conclusions

This chapter has reviewed sampling, flow measurement and analysis, together with the reasons for making a representative and accurate job of each, and interpreting the results. It would certainly be futile to construct a treatment plant, at considerable capital expense, and then expect consistent performance without monitoring and controlled feedback.

Wastewater influent quality changes as production processes or the connected population do, and trend analysis of a reliable databank of historical results provides all the information to make decisions about expanding or modifying treatment facilities. It will soon become a legal requirement in many countries to monitor treatment plant performance and a condition of Discharge Consent issue.

Where manpower resources do not allow an adequate monitoring programme to be maintained, an increasing number of environmental consultants offer this service and can provide experienced interpretation and reporting. The author has been involved in a number of these where considerable savings have been made to water and raw material consumption and a number of expensive modifications shown to be unnecessary or worth a rethink; and all for the price of a well-designed survey and appropriate analysis.

8.13 Companies and other organisations

ABB Kent—Taylor Ltd, Howard Road, Eaton Socon, St Neots, Cambridgeshire PE19 3EU.
ELE International Ltd, Eastman Way, Hemel Hempstead, Hertfordshire HP2 7HB.
Green World Instruments Ltd, Fir Lodge, 98 Holland Road, Maidstone, Kent ME14 1UT (UK agents for ISCO Environmental Division, 531 Westgate Boulevard, Lincoln NE 68528-1586, USA).
Her Majesty's Stationery Office (HMSO), PO Box 276, London SW8 5DT UK. Also from HMSO bookshops.
Ionics UK Ltd, Unit 3, Mercury Way, Mercury Park Estate, Urmston, Manchester M41 7LY.
Mettler Toledo Ltd, 64 Boston Road, Beaumont Leys, Leicester LE4 2ZW.

Montec International Ltd, Pacific Way, Salford, Manchester M5 2DL.
National Measurement Accreditation Service (NAMAS), Building 202, National Physical Laboratory, Teddington, Middlesex TW11 0LW.
Pollution Prevention Monitoring Ltd, The Bourne Enterprise Centre, Borough Green, Sevenoaks, Kent TN15 8DG.
Warren Jones Engineering Ltd, 120–124 Churchill Road, Bicester, Oxfordshire OX6 7XD.
Whatman Labsales Ltd. St Leonards Road, Maidstone, Kent ME16 0LS.

Chapter 9
Case Histories

9.1 Introduction

This chapter gives brief accounts of a few of the wastewater plant rebuilds, modifications and upgrades with which the author has been involved, including industrial effluent problems solved without recourse to a treatment plant. Industrial effluent and sewage treatment problems often have common themes and the reader might therefore find general perusal of this chapter useful. It is hoped that the reader will be able to identify some similar features to plant they are operating or planning and the accounts provide direction if not necessarily the definitive answer to problems.

All wastewater plants are unique; nevertheless, some basic self-help principles apply if founded on reliable data. Once again, the emphasis is on proper sampling, survey and analysis of the effluent before decisions are made. If you do not have the time or resources for these activities, employ a consultant or specialist to carry them out: £20 000 spent on a survey may save £250 000 on an unnecessary or over-designed treatment facility. The survey should look at where effluent is being generated, its components, why they are there and methods to reduce both volume and strength. The possibility of recycling cooling water, particularly that to maintain gland seals in pumps, should be examined.

A recurring theme on sites where treatment is proposed for the first time is the volume of rainwater discharging to the sewer with contaminated wastewater. Its removal often halves the volume of effluent needing treatment. Other issues come to light when drainage is segregated: the state of the drainage system, storage and correct bunding of chemicals, COSHH and safe working practices and where spillages are likely to occur.

The rainwater, prolific in many areas, might be profitably viewed as a resource in some instances and used as second grade water where minor contamination is unimportant. Virtually all these aspects are present in a multimillion pound proposal with which the author has been involved through a client at the time of preparing this text. In common with other organic dischargers, a trio of food manufacturers is faced with the option of either treating their effluent and discharging to river or facing a 100-fold increase in effluent discharge costs over the next 18 months. Situations like this are becoming increasingly common in the UK and Europe in the 1990s as the Water PLCs are required to install new treatment facilities or upgrade existing ones to conform to the EC Urban Wastewater Directive (91/271/EEC) (1991). Clearly, capital and operating costs will be passed on.

In the particular cases mentioned above, and as a result of complexities regarding tenancy agreements, the three food manufacturers decided to go their own ways, and in the medium term only one will be constructing a treatment plant. This company will be discharging direct to the river and a high quality effluent will be demanded; the onus will be placed on the firm to acquire rapidly the necessary operating skills. All site surface water is to be redirected before the effluent plant designs are drawn up. The pay-back period is estimated at 3.5 years.

The second discharger has undertaken a radical review of effluent generation in the factory and already halved the discharge volume to foul sewer by recycling 90% of the cooling water. This company also proposes to catch enough of the 2000 mm of rainwater that falls on its manufacturing site annually to use for general washing down of areas not connected with food production. It is estimated that these actions will reduce the potential trade effluent bill by 80% and the increase will therefore be reduced by 20 times (£1000 per year instead of £20 000). Doubts over the long-term presence of the company on the site has led to a situation in which a relatively short-term view of effluent treatment has been taken.

The third manufacturer has already started to make significant alterations to the drainage system in its factory area. It is apparent that considerable infiltration from ground- and rainwater is occurring and the complexities of the site have resulted in intermingling of this company's discharge and that of the second manufacturer mentioned above – whereas they wish these to be quite separate!

Longer-term tankering off-site of the strongest discharges is a possibility as the close presence of housing mitigates against a roughing filter plant although one of the enclosed package units might be considered. The effluent has a severe nutrient imbalance and so any treatment unit will require flow balancing, chemical addition and some supervision.

Perhaps the two most significant factors in all three cases, brought into sharp focus by the developing strategy on trade effluent charges, are:

(1) How much water wastage has occurred historically; all three factories have borehole supplies at very cheap rates that only require on-site chlorination and storage.
(2) The high percentage of rainfall in the present discharges to foul sewer – in one case 90%. The historical pumping costs for this, mostly incurred by the local council and later the Water PLC, must have been considerable.

242 *Sewage and Industrial Effluent Treatment*

9.2 Case history no. 1

9.2.1 Background

A dairy principally involved with bottling milk wished to improve the effluent quality from its on-site 20-year-old treatment plant discharging to a river, anticipating the setting of higher standards and an ammonia limit.

The facility treats an alkaline wastewater with CODs and BODs of 1500+ mg/l, low semi-colloidal solids around 500 mg/l and with little nitrogen and phosphorus − a typical dairy waste. The main treatment unit, a high rate biotower, has been supplemented in recent years by a DAF plant. The original Discharge Consent limits of 400 SS and 200 BOD mg/l were likely to become 60 SS/40 BOD/5 NH_3N mg/l within three years.

There is little spare room on this site and a major trunk road has been built very close to the high-rate biotower (Figure 8.1). Figure 8.2 shows a panoramic view of the plant before modification. Situated in a development zone, modi-

Figure 9.1 High-rate biotower and proximity of roadway.

Figure 9.2 Panoramic view of dairy effluent treatment plant.

fications would not be difficult to approve and planning permission easily obtained without delay. The plant is fairly simple in layout and quite easy to maintain, but the flat site results in a lot of pumping initially and between stages. Historical problems had centred around pump failures and poor pH dosing control causing alkaline discharges to damage the biotower media. Regular internal audit sampling and analysis of the effluent has been conducted and provided a useful database; the NRA took regular samples and for the most part these complied with the very relaxed Consent values.

9.2.2 Process description

Alkaline washings and discarded milk from the bottling plant enters a low level sump and is pumped to a balancing tank where air sparging prevents settlement and anaerobic conditions. Two forward feed pumps convey the wastewater to a pH dosing chamber where sulphuric acid is added to reduce the pH from 10–12 to 7–8 (Figure 8.3). Polyelectrolyte and ferrous sulphate are then added by metering pumps and the effluent enters a DAF plant where some 80%–90% of the fat and protein is skimmed off the surface (this being a later addition to the plant).

The effluent gravitates to a low level sump and mixes with biotower effluent, urea solution and phosphoric acid (nutrients to provide N and P via fixed rate pumps). It is then pumped to the biotower, a plastic media, asbestos-clad structure 10 m tall. Effluent drains via a recirculation sump to two settlement tanks. These are desludged four times/day; the final effluent flow rate is measured before discharge to the river being typically 600 m^3/day.

As the flow rate from the bottling plant is negligible between midnight and

Figure 9.3 pH correction and chemical mixing tank.

6 AM and nil from Saturday evening until Sunday afternoon, the forward feed pumps are manually throttled at the end of the day and on Saturday late morning so that the plant does not in theory drain the balancing tank entirely. Nevertheless, erratic flows do occur and the recirculation ratio to keep the biotower wet often exceeds 8:1. The sludges from the settlement tanks and the DAF unit are tankered off-site without any treatment and soil-injected.

9.2.3 Preliminary work

Despite the historical final effluent data, little was known about the biotower performance, pH control, nutrient addition and flow rates and a full ten-day survey instigated. Influent and effluent and biotower feed and effluent were measured by time proportional auto samplers and samples of sludge taken manually. About 200 samples were taken in total.

Effluent flow rate was measured independently and a discrepancy of 20% found against the installed equipment. From the survey, the main findings were:

(1) Erratic flows through the plant leading to very high recirculation rates and pumping costs.
(2) The DAF plant was removing 85% of the load to the plant, leaving the high-rate biotower with an influent similar to settled sewage and too weak

for such a unit with 90% voidage. Consequently, BOD removal rates were less than 10%.
(3) Some collapse of the media in the base of the biotower was evident from the increase in ammonia through the unit and the reduction of residual ferrous sulphate to sulphide, which generated hydrogen sulphide and caused poor solids removal in the settlement tanks.
(4) The settlement tanks were undersized and the upward flow rate too high during the day to effect good settlement.
(5) Nutrient dosing rates were not flow-related and the biomass in the biotower was being periodically starved of N and P.
(6) The pH control system was set wrongly and over-reactive; the pH of the biotower feed was too high and rapidly changed.

9.2.4 Solutions

The client agreed to a programme of modifications involving:

(1) Trebling the balancing facilities by adding another tank. This allowed five days' worth of flow to be pumped through the plant over seven days and gave extra reserve for a proposed 25% increase in production in the near future. New variable speed pumps were installed. Because of the cramped site, the tank was located on the other side of the main road flyover and levelled so that it would fill and empty at the same rate as the original balancing tank. Isolation valves were provided.
(2) pH control was improved by providing better and more responsive control loops. The overflow direct to the biotower was re-routed to the biotower feed sump which was considerably enlarged and provided with acid dosing and a control electrode.
(3) The biotower media was replaced with a closer packed material designed for conventional BOD loadings and flow rates and the tower enlarged. The asbestos cladding was replaced and disposed off-site by a specialist contractor.
(4) An extra settlement tank was provided and pipework included so that a second biofilter could be added if necessary.
(5) Nutrient dosing was coupled to the final effluent flow recorder.
(6) The sludge from the DAF plant was sent for dewatering trials and different conditioners tried out. The thought here was to convert a £50 000/year disposal cost to a potentially saleable animal feed product high in fat and protein. At worst, dewatering would permit disposal by skip to landfill, currently about one-third the cost of liquid tankering.

The end benefits achieved have been:

(1) The effluent from the plant achieves the new standard of 60 SS and 40 mg/l BOD 97% of the time. Winter performance is noticeably better as the

biotower acts less like a wind tunnel than the previous high voidage structure.
(2) A 30% reduction in pumping costs. Flow rate through the plant is now nearly constant because of the large increase in balancing tank capacity and the amount of recirculation required is minimal.
(3) Sludge tankering costs have been reduced but the final savings are uncertain at present due to the difficulty of finding a local outlet for the dried sludge. However, skip disposal costs have proved to be 30% of the original tankering costs.
(4) There are fewer smell complaints from the neighbours about the plant in summer.
(5) Chemical consumption has dropped due to better control, and the extra balancing facilities have introduced a beneficial damping out of sudden pH variations.

One of the main features of this plant was that the installation of the DAF unit a few years previously had rendered the high rate biotower virtually superfluous in terms of roughing a high strength waste. Hence its conversion to a conventional biofilter to achieve something approaching a good quality final effluent.

The problems of sludge disposal are also emphasised here for the DAF plant skims off a very acceptable animal feed supplement high in fat and protein, but an outlet has proved elusive. Tankering to land is not a long-term solution as costs rise and outlets diminish.

Poor pH, nutrient and chemical coagulant control are commonly found on industrial effluent plant. These must be flow related via good quality metering pumps and electrodes and pump dispensing rates regularly checked.

9.3 Case history no. 2

9.3.1 Background

A manufacturer of domestic items plating chromium, nickel, and zinc and connected to the foul sewer appeared to suffer sporadic losses of settled metal hydroxide sludge from the settlement tank. This came to light originally when the staff at the local receiving sewage works noticed a considerable drop in the metal levels of the sludge during holiday periods when the factory was closed. The chromium levels of samples frequently exceeded the 5 mg/l limit in the trade effluent Consent, particularly when the plating shop had a heavy workload.

9.3.2 Process description

A small package treatment plant dosed sulphur dioxide gas to reduce the chromium to Cr^{3+} on a separate line after the necessary pH reduction (see Chapter 3). Along with the zinc and nickel, this was then precipitated as the hydroxide by adding sodium hydroxide solution and maintaining the pH between 8.5 and 9.2. Cyanide was converted to cyanate by sodium hypochlorite solution.

Final settlement was confined to a large, rectangular tank with sloping floor and scum boards but no baffles. Sensing electrodes were regularly serviced; the treatment plant contained a lot of 240 V solenoid valves directly dosing the chemicals.

9.3.3 Preliminary work

This was confined to the examination of many sample results taken by the trade effluent inspector and related correspondence.

9.3.4 Solutions

An on-site meeting was held in order that a critical view could be taken of the facilities, and the major problem was solved in five minutes – literally. In an attempt to improve flow baffling in the settlement tank, the scum boards had been made deeper and lowered into the tank further. Consequently, flow surges were passing under the bottom of the boards and stirring up the settled hydroxide sludge, a light flocculant material that always needs quiescent conditions.

Cutting the scum boards in half and raising them so that immersion was not more than 0.2 m (the tank was about 2 m sloping to 2.5 m) provided an instant cure to hydroxide solids carryover. GRP inlet and outlet flow baffles were fitted to the tank to counter surges and a small upstream balancing tank proposed. Later, optical sensing of the sludge level in the tank coupled to automatic desludging was installed.

A second prefabricated sludge holding tank was purchased which allowed considerable dewatering of the sludge, such that tankering trips were halved in number. Ultimately, pressing to produce a cake might be desirable and reduce transport costs even more, although the best state for metal reclamation may dictate retaining the sludge as a concentrated liquid.

Although the metal levels in the effluent generally dropped and the zinc equivalent of the sludge at the local works dropped by nearly three times, the reduction was less than predicted and chromium remained erratic and high.

Two further problems were identified:

(1) The sulphur dioxide gas was being poorly mixed and metered into the Chromium line effecting variable Cr^{6+} to Cr^{3+} reduction. Because of handling problems with gas cylinders, the firm decided to use sodium metabisulphite instead.

(2) The 240 V solenoid valves admitting treatment chemicals 'chattered' when the plating shop was in heavy use because the plating bath current transformer load caused a considerable voltage drop in the building – down to 210 V. A rewiring of the treatment plant to a separate supply cured the problem of oscillating valves and variable chemical dosing.

While the solutions to this case may in retrospect seem facile, they serve to underline the benefit of an outsider who hopefully spots the obvious which the on-site employee, engrossed in production, may miss.

The longer-term consequences of recommending a 'quick-fix' always need thinking through. Although historical, this case underlines the simplicity of some problems and the need to address final disposal of end waste products with long-term economic solutions.

9.4 Case history no. 3

9.4.1 Background

A major food manufacturer was requested by the Water PLC to install a flow recorder and proportional sampler so that every week, a 24-hour flow proportional sample could be taken for analysis. Three months' worth of values produced an average for charging purposes using the formula described in Section 7.9 of Chapter 7. It was soon realised that under this new sampling arrangement, the company had incurred an eight-fold increase in its trade effluent bill, previous samples being spots taken during the working day by the company and the Water PLC inspector.

No treatment of effluent takes place on the site which is in the middle of a housing estate. Six separate discharges occur within the factory, one of which is almost entirely domestic sewage; another is cooling water. Production areas discharge separately to the factory drains, the effluent from each being very variable in strength and volume.

9.4.2 Preliminary work

The client requested a review of options available with the primary desire of minimising effluent disposal costs. An eight-day survey was undertaken, this being the time span of a production cycle. As the budget was limited, automatic time proportional samples were taken from two principal areas for eight days and the discharge from the dairy intensively monitored for 24 hours. Other areas were manually sampled and an automatic sampler moved about the site to different locations to effect short 'blitz' sampling. The installed flow recorder and composite sampler were also used and as it proved impossible to measure flows accurately at a lot of the discharge points, provided the basis for some realistic flow proportioning estimates for the separate discharges.

Samples were analysed by an external National Measurement Accreditation Service (NAMAS) accredited laboratory for the usual parameters of Total and Settled COD, SS, BOD, pH, Settleable Solids, NH_3N. Carbohydrate and grease and oil were measured in some samples. The suites were chosen to allow cost to be calculated for the individual discharges using estimated flows and also assess treatability of each flow.

9.4.3 Solutions

The analytical results indicated a wide variation in strength between the various components of the total factory discharge. The average composite from the whole site had an organic strength ten times that of domestic sewage, with the main production flow, 60% of the total, being five times the organic strength.

One particular discharge was only 6% of the flow from the factory site but 75% of the total organic load, and yielded values of 150 000 mg/l COD (Settled) and 30 000 mg/l Settleable Solids with 60 000 mg/l carbohydrate. Daily volumes of this discharge varied with the number of wash-ups and it became obvious from observation that different working practices contributed to the number of wash-ups.

Many samples from all production areas were highly putrescible but in general the solids did not settle well, contributing to a high soluble BOD. The options available were considered to be:

(1) Treat the effluent with an on-site facility and discharge to the local stream.

This was discounted on the grounds that a considerable amount of on-site expertise would be necessary to treat an effluent high in organic strength and carbohydrate, the latter proving problematic operationally to an activated sludge plant (which would probably be required to minimise odour). It is doubtful whether planning permission would be granted for such a plant so close to housing. Underground and package plants were discounted for similar reasons.

The capital cost of this option exceeded £0.7 million and the company viewed the long-term future of the production site as uncertain.

(2) Partially treat the effluent, either all the discharges or selected ones, and discharge to the sewer.

Either high-rate biofilters or primary settlement were contemplated. Both were discounted, as although reducing the strength would reduce trade effluent charges, the analysis has indicated poor settleability. High-rate biofilters are often very odorous, particularly treating high carbohydrate wastes and, again, planning permission was not likely to be granted. There would still be highly reactive sludges to store and dispose of. Running costs would probably exceed savings and again capital costings were unattractive.

(3) Isolate the strongest discharge and tanker it off site separately. At the same time, examine the work practices generating it and review water consumption to minimise waste in this production area.

This option was estimated to reduce the trade effluent charge/day from £850 to £260, while the cost of tankering out this 18 m^3 discharge/day was about £270, a saving of £320.

Disadvantages with this option include providing a tank of 50 m^3 to store three days' production of a highly putrescible liquid and reliance on a reliable tankering service.

As the material is a controlled waste (and the EPA 1990 Section 34 requires the company to exercise Duty of Care), it can only be disposed of to a licenced site, a sewage works 60 km away. Transport costs are therefore high but the strength of the waste, which has to be tested regularly, is less of a factor in this case.

At the time of writing, the company has pursued Option (3) and is further experimenting with dewatering the sludge produced to reduce tankering, returning the relatively clear liquor to the sewer. Longer-term thermal drying along with other food-based waste from the factory is proposed. There is a steadily growing market for the sale of this type of reclaimed animal feed product to local farmers.

Savings have been effected in this case without treatment, merely by separation and re-routeing the strongest discharge, which in time may prove to be a saleable commodity. With the present interest by the Water PLCs in reducing their trade effluent loadings to save on capital expenditure for sewage works extensions, the industrial organic dischargers are becoming prime targets and many have received un-welcome increases in trade effluent costs in hard-pressed economic times. Here, a solution was achieved without treatment, by separating the waste streams and looking at water use and work practices. There are many more industrial situations where this is the best route to take.

9.5 Case history no. 4

9.5.1 Background

Operators of small private sewage works are periodically likely to be faced with a major repair despite planned maintenance. The predicted solution of abandoning the works and connecting to foul sewer may not, however, prove to be the best one.

In this example, the main rotor or drive shaft of a rotating biological contactor (RBC) had collapsed and with it both bearings after 12 years of use. This is not uncommon with units of this vintage, as the weight of media which build up on the discs and becomes unevenly distributed during a loss of rotation places severe mechanical strain on shaft, bearings and motor drive.

The repairs, including changes to the GRP casing, a new drive chain and motor were estimated to cost £16 000. Some of this was accounted for by the cranage costs to remove and replace the shaft.

Figure 8.4 shows a cross-section through an RBC unit designed to treat domestic sewage from 70 people. The plant served a small conference centre and an average population of 90 for which it was generously sized. The

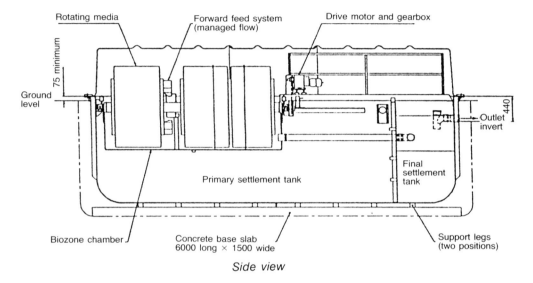

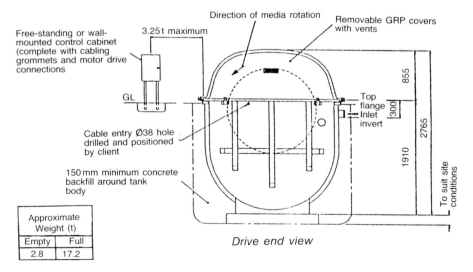

Figure 9.4 Cross-section through an RBC of 70 pe size. (All dimensions are in mm.) (Courtesy: Klargester Environmental Engineering Ltd.)

collecting sewers suffered infiltration and flow rates in heavy rain sometimes flooded the plant briefly. The discharge to a local stream was semi-nitrified and augmented the flow in drought conditions. Consent limits in mg/l were 75 SS and 50 BOD.

Concerned by the magnitude of the repair bill, a lack of skilled on-site labour and the possibility of other failures to an ageing plant, the client wished to review other options for sewage treatment or disposal.

9.5.2 Preliminary work

A site visit was made to assess the general condition of the plant which proved structurally sound, well maintained by external contractors and well installed. Several local people were contacted and the general lie of the land assessed with respect to the nearest public foul sewer. Discussions took place with the Water PLC and the local river regulating body.

9.5.3 Solutions

Solutions in such cases are bound to be unique to each location and there may be no practical alternatives to continuing sewage treatment on site. Four options were identified for the client in this case, but on estimating costs, only one proved practical in the medium term as described below.

(1) *Abandon sewage treatment and connect to the local sewers.* These were located 1.5 km away in a village about 150 m lower than the site. The agricultural land in between undulated and the prospect of laying a sewer on a gently falling gradient would have required tunnelling in at least two places. Both the local stream and minor road described a meandering course to the village.

No detailed civil costs were calculated as a rough estimate put this option at over £200 000, without any allowance for an intermediate pumping station or connection costs. Annual sewerage charges would cost £2500. Savings of £550/year would be made in running and maintaining the RBC plant.

In initial consultation and before a site visit, this option had looked attractive but in this case would have been prohibitively expensive if the sewer connection exceeded 500 m. Moral – do not make even preliminary judgements from the office!

(2) *Refurbish the present RBC sewage treatment plant.* After competitive estimates had been received and repair costs dropped to £13 000, this became the much preferred option. Other favourable factors that evolved in the discussions included no intention by the River Authority of setting higher discharge standards in the next five years or the client extending conference facilities. Thus, when repaired, the unit should prove adequate for at least another ten years.

A marginal environmental advantage was the maintenance of flow in the local chalk stream during drought as a result of the plant discharge of about 20 m^3/day.

Costs of £13 000 to repair, £550/year to run and maintain and £1150/year to discharge were estimated. It was recommended to the client that a loss of rotation sensor and high level water alarm be fitted to the refurbished unit with the annunciator in the main office. They were also advised to thoroughly de-sludge the RBC regularly and so prevent excessive soluble BOD and consequently heavy biomass build-up straining the shaft and drive.

(3) *Abandon the RBC for an alternative which would produce a better effluent and cost less to run.* If tighter discharge limits were soon on the way and energy or maintenance costs high, this would be a sensible choice in many cases. Here, it would simply 'over egg the pudding' and exceed the client's capital budget. If an ammonia standard was applied in five years' time, it would be quite feasible to add a nitrifying filter to the RBC retrospectively.

The alternative choice, costed at £35 000 for this location, would involve pumping to a high level primary tank with influent screening. Gravity-fed irrigation of a close packed percolating filter would provide the biotreatment section, draining to the shell of the RBC unit converted to a settlement tank. Some dismantling costs for the RBC were included in the estimate. A 30 SS/20 BOD or better nitrified effluent would be expected.

The new plant would be more visually intrusive and might be an odour source unlike the enclosed RBC, although situated a long way from houses. The intermittent pump would use less energy than the continuous drive of the RBC and the plant would comprise package units with few mechanical parts. Desludging requirements would remain but running costs/year should therefore be less. Effluent discharge costs/year would stay the same.

An area of major uncertainty where new structures are to be placed on a contemporary site are ground conditions and loading rates, i.e. the ability to support heavy tanks without building substantial foundations.

(4) *Abandon sewage treatment and tanker the sewage away.* This option would be favourable in some instances and it might be the only choice. Mitigating against it in this case was the cost of tankering at £8/m^3, even with a heavy discount, and the lack of substantial storage necessitating almost daily visits. Because of infiltration, the client would also be paying a large sum of money in a wet winter to have polluted rainwater taken down the road. In fact, annual disposal costs equalled option (3) above, a new plant. The effluent discharge costs of £1150 would be saved, however.

If a septic tank was operated instead, the top liquor could in theory be discharged to a soakaway with Consent and considerably reduce tankering costs. At this location and because of the chalk substrata, the River

Authority rarely permitted this arangement and this alternative was not costed.

This case serves to illustrate the range of individual factors that must be considered and the unique aspects of each location influencing costs. As a general rule, it is probably better to persevere with treatment facilities (unless catastrophically damaged) than place operational reliance and security with third parties where costs and the weather can change rapidly.

9.6 Conclusions

To round off these case histories, the author includes two delicious and lighthearted résumés of how not to effect good sewage treatment, but perhaps the perfect (immoral) solution to effluent quality problems.

The first concerns an impeccably maintained small sewage works, almost the perfect secluded picnic site. Every year in June, the whole works was refurbished, all equipment dismantled and rebuilt, everything painted and the grass manicured to resemble a billiard table. Unfortunately, the works failed the Consent standards of 30/20 for about three months after this activity and probably because all the biofilter media was removed, steam-cleaned and then replaced, not necessarily in the order in which it had been removed.

The second was a tiny works which by all accounts always produced a fabulous effluent where most analytical parameters were always measured virtually at the detection limit of the method – until the lone character working there retired and took with him one of the most endearing operational practices imaginable. On instruction from on high and the night before the effluent inspector was due, the humus tank was drained and refilled with tap water, a lengthy process taking 14 hours. This was in the days when watches could be set by the arrival of the local river or effluent inspector.

Appendix 1
Glossary of Terms
(Courtesy: CIWEM)

activated sludge the microbial mass of micro-organisms (mainly protozoa, bacteria) and inert material generated by the continuous aeration of sewage.

aerobic A condition in which oxygen as gas is freely available and used in the free form.

alkalinity the ability of water to neutralise acids due to the presence of bicarbonate and carbonate and sometimes hydroxide. Expressed as mg/l calcium carbonate.

allylthiourea a chemical commonly added to the dilution water of the BOD test to inhibit the action and oxygen demand of nitrifying organisms present. The carbonaceous BOD is thus obtained.

anaerobic the absence of oxygen in either the free form or as nitrate/nitrite/phosphate.

analytical quality control a system of internal laboratory auditing using standards, blanks and spiked samples to validate the accuracy of sample analytical values.

anion a negatively charged ion, e.g. phosphate, nitrate, chloride, sulphate.

anoxic the absence of free oxygen but where oxygen is available as nitrate/nitrite/phosphate (but not sulphate).

autotrophic (bacteria) deriving energy from inorganic reactions and carbon from carbon dioxide or bicarbonate. Nitrifying types are examples.

biomass the mass of organisms comprising the activated sludge or film growth in a treatment system, including dead material and debris.

bulking a condition where activated sludge occupies excessive volume and does not settle readily, causing solids loss from settlement tanks. Often caused by filamentous organisms.

calorific value the number of joules of heat derived from complete combustion of a fuel (sludge or gas) expressed as J/g, MJ/kg, etc.

carbonaceous oxidation the biochemical oxidation of carbonaceous matter to carbon dioxide.

cation a positively charged ion, e.g. sodium, calcium, magnesium.

clarifier a tank designed to reduce the turbidity or solids content of wastewater by settlement or chemical precipitation.

classifier the part of grit removing equipment designed to wash the grit and remove it by rake or centrifugal action.

combined heat and power an engine/generator system producing electricity and usable heat from the combustion of (sludge) gas.

comminutor a hollow drum with horizontal slots, teeth and combs which shreds solids in sewage at the works inlet. Largely superseded by fine screens.

Consent a legal document permitting discharge of effluent to sewer to watercourse and usually containing limits regarding flow rate, total volume and composition.

controlled water territorial, coastal, inland and ground water as defined by the UK Water Resources Act 1991, Section 104, with powers to prevent and control pollution.

copperas ferrous sulphate $FeSO_4 \cdot 7H_2O$ used in solution form and often as a coagulant in sludge treatment.

denitrification the reducing of nitrate and nitrite, often by bacterial action, to nitrogen gas.

Detritor the trade name of a grit collector and cleaning channel.

detritus inorganic debris often with some organic material (sewage treatment); otherwise a general term for dead and decomposing biological matter.

dry weather flow the average flow to a works during a period of nil rainfall – usually over more than seven days. Seasonal variations are also taken in to account.

endogenous (growth or respiration) the negative phase of microbe growth where, due to nutrient depletion or other adverse circumstances, the cell contents are used to effect respiration. Dead cells may also be used, leading to a decline in population numbers.

eutrophication the enrichment of natural waters by compounds of nitrogen and phosphorus.

facultative as applied to bacteria, able to live in either state, i.e. anaerobic or aerobic conditions.

flocculation coagulation of fine and colloidal solids to form larger particles that will settle more readily. Chemicals or slow stirring are often used.

flotation the removal of solids from wastewater by inducing them to float rather than settle, using fine air bubbles; often used for fatty materials whose density is very close to or less than one.

F/M ratio the food-to-mass ratio in a system; in activated sludge plant, the ratio of the BOD load (in kg/day) applied to the total volatile or mixed liquor suspended solids already in the system.

gas liquid chromatography a separation technique for identifying components in a mixture. This method is particularly used for identifying volatile organic compounds with boiling points below about 400°C, e.g. solvents, and is extremely sensitive – levels of $<10^{-15}$ are possible.

heavy metals copper, zinc, chromium, cadmium, lead and nickel are the usual six elements associated with this term. Commonly used in the metal finishing industry and toxic to wastewater treatment plant.

ion selective electrode a measuring device that develops a small potential difference proportional to the activity or concentration of a particular ion, e.g. Na^{2+}, NO_3^-, F^-. Widely used in continuous monitoring systems, electrodes have a high degree of selectivity and are quite robust; they can, however, suffer from interferences from other ions and surface active materials.

nomograph a graph in which the value of a variable can be found when the values of two other variables are known. Three parallel vertical lines are used, the outer two being calibrated with the two known variables and the middle one with the unknown. A straight line joining the two outer lines cuts the third at the value sought.

obligate as applied to bacteria, able to live in one state, i.e. aerobic or anaerobic, and unlikely to survive in the other.

polyelectrolyte a general term for chemical conditioners widely used in sludge thickening and dewatering and water clarification. Most are high molecular weight organic compounds with a number of ionisable groups and are dosed at mg/l levels, unlike inorganic chemicals such as lime and copperas.

population equivalent defined as:

$$\frac{BOD\ (mg/l) \times flow\ (m^3/day)}{0.060 \times 10^3}$$

This is a term used to compare the strength of industrial wastewater with domestic sewage and based on the premise of 0.060 kg BOD load per capita per day.

Red List the UK version of the Black List and List 1 substances, as defined in 76/464/EEC (the Dangerous Substances Directive) and others. These are the most dangerous substances to the water environment by nature of their persistence and toxicity.

sewage the polluted wastewater derived from domestic and industrial activity.

sewerage the system of pipes and related structures conveying sewage to a treatment plant or discharge point.

sloughing excessive discharge of solids comprising the film of micro-organisms growing within a biofilter; seasonal activity occurring in the UK in spring and autumn when ambient temperatures change.

sludge age the number of days it would take to waste the total mass of sludge from an aeration plant under stable operating conditions.

sludge volume index a measure of sludge settleability. Defined as the volume in millilitres occupied by 1 g of activated sludge after 30 minutes' settlement.

$$SVI = \frac{\text{settled volume of sludge (ml)}}{\text{mixed liquor suspended solids (mg/l)}}$$

To emulate the dynamic conditions in a settlement tank, the **stirred sludge volume index** is often used instead in order to provide a more realistic result.

surface water rainwater run-off from paved areas, roads, buildings and land.

trade effluent liquid produced in the course of industrial production activities which may be polluting and the subject of Discharge Consent. Defined in the Water Industry Act 1991, Section 141 (1).

volatile acids acetic, propionic, butyric, valeric acids, etc., produced during sludge liquefaction, anaerobic digestion or the breakdown of most organic material in anaerobic or semi-anaerobic conditions.

volatile matter the proportion of material (in sludge) lost after heating to 600°C and equivalent to the organic content.

Appendix 2
Typical Sewage and Industrial Effluent Plant Layouts

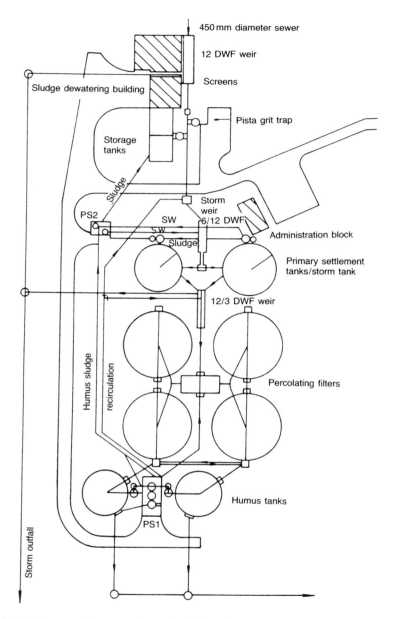

PS1 Includes pumping sumps for recirculation, sludge and alternative double filtration system.
PS2 Includes pumping sumps for sludge and storm water recirculation.
SW = stormwater.

Layout of a percolating filter STW designed to treat 900 m^3 DWF/day of domestic sewage. (Courtesy: CIWEM.)

Typical Sewage and Industrial Effluent Plant Layouts

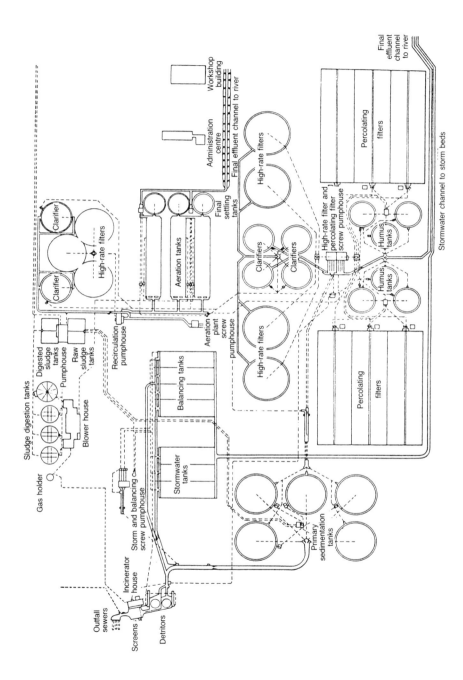

A sewage works treating 40 000 m³/day of domestic sewage and a mixture of trade effluents with activated sludge, pre-treatment high-rate filters and conventional-rate filtration, with sludge digestion and power generation. (Courtesy: CIWEM.)

Appendix 2

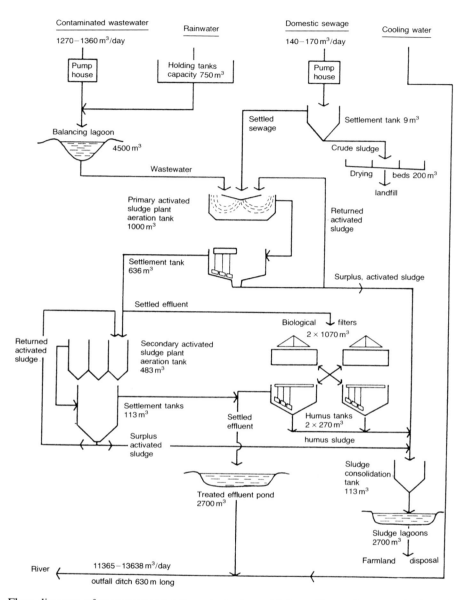

Flow diagram of wastewater treatment plant combining effluent from synthetic resin manufacture and domestic sewage; a large, open site with both activated sludge and filters. (Courtesy: CIWEM.)

Typical Sewage and Industrial Effluent Plant Layouts 263

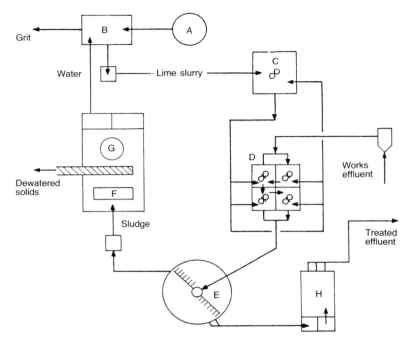

A Quick lime storage tank.
B Lime slaker.
C Lime slurry recirculation tank.
D Four compartment reactor tank.
E Thickener.
F VC filter.
G Rotary vacuum filter.
H Treated effluent tank.

Industrial effluent treatment plant removing toxic metals from dyestuffs production wastewater by precipitation with lime. (Courtesy: CIWEM.)

Appendix 3
Imperial/Metric (SI) Conversion Factors

Multiply imperial unit by figure in multiplier column to obtain metric (SI) equivalent; multiply metric (SI) unit by reciprocal to obtain imperial equivalent.

Imperial unit	Multiplier	Metric unit	Reciprocal
acre	0.4047	hectare	2.471
Btu	1.055	kJ	0.9478
Btu/ft^2 h	3.155	J/m^2 s	0.3170
Btu/ft^3	37.26	kJ/m^3	0.0268
Btu/gal	0.2321	kJ/l	4.31
Btu/h	0.2931	W(J/h)	3.412
Btu/lb	2.326	kJ/kg	0.43
°F	0.5555 (°F − 32)	°C	(1.8 × °C) + 32
ft	0.3048	m	3.281
ft/min	5.08	mm/s	0.1969
ft/s	0.3048	m/s	3.281
ft^2	0.0929	m^2	10.76
ft^2/ft^3	3.28	m^2/m^3	0.305
ft^3	0.02832	m^3	35.31
ft^3/ft^2	0.3048	m^3/m^2	3.281
ft^3/gal	6.23	m^3/m^3	0.1605
ft^3/lb	0.0624	m^3/kg	16.1
ft^3/mil gal	0.00623	m^3/10^3 m^3	160.51
ft^3/min	0.000472	m^3/s	2119
ft^3/min ft	0.00155	m^3/s m	645.86
ft^3/min ft^2 plan area	0.00508	m^3/s m^2 plan area	196.85
ft^3/s (cusec)	0.02832	m^3/s (cumec)	35.31
gal	4.546	litre	0.220
gal	0.004546	m^3	220
gal/ft	0.0149	m^3/m	67.114
gal/ft^2	0.049	m^3/m^2	20.408
gal/ft^2 h	1.176	m^3/m^2 d	0.850
gal/ft^2 min	70.56	m^3/m^2 d	0.0142
gal/min	0.0758	l/s	13.192
gal/yd^2	0.005437	m^3/m^2	183.9
gal/yd^3	0.005946	m^3/m^3	168.2
hp	0.7457	kW	1.341
hp/mil gal	0.164	W/m^3	6.098
in	25.40	mm	0.03937
in^2	645.2	mm^2	0.00155

in^3	16.39×10^3	mm^3	0.061×10^{-3}
lb	0.4536	kg	2.205
lb/ft^2	4.881	kg/m^2	0.2049
lb/ft^3	16.03	kg/m^3	0.06243
$lb/10^3\, ft^3$	0.016	kg/m^3	62.5
lb/in^2	0.06895	bar	14.503
lb/yd^3	0.5932	kg/m^3	1.686
mil gal	4546	m^3	0.220
mil gal/acre	1.123	m^3/m^2	0.890
mile	1.609	km	0.6214
ton	1.016	tonne	0.9842
yd	0.9144	m	1.094
yd^2	0.8361	m^2	1.196
yd^3	0.7646	m^3	1.308

Common Wastewater Imperial/Metric (SI) Parameters and Conversion Factors

Parameter	Imperial unit	SI unit	Conversion factor
Rainfall	inches	mm	25.4
Sewer (diameter)	inches	mm	25.4
Sewage			
flow	gal/min	l/s	0.0758
	gal/d	m^3/d	0.004546
	mil gal/d	$10^3/m^3/d$	4.546
BOD per capita	lb BOD/hd/d	kg BOD/hd/d	0.4536
Pumping rate	gal/min	l/s	0.0758
Preliminary treatment			
Screenings	ft^3/mil gal	$m^3/10^3\, m^3$	0.00623
Grit removal:			
velocity	ft/s	m/s	0.3048
volume of grit	ft^3/mil gal	$m^3/10^3\, m^3$	0.00623
air	ft^3/min ft	m^3/s m	0.00155
Primary treatment			
Capacity	gal/hd	l/hd	4.546
Surface loading	$gal/ft^2/d$	$m^3/m^2/d$	0.049
Upward-flow rate	ft/h	m/h	0.3048
Weir loading	gal/ft/d	m^3/m/d	0.0149
Scraper speed	ft/min	mm/s	5.08

Parameter	Imperial unit	SI unit	Conversion factor
Secondary treatment			
Biological filters:			
loading:			
hydraulic	gal/yd^3/d	m^3/m^3/d	0.005946
superficial	gal/yd^2/d	m^3/m^2/d	0.005437
BOD	lb BOD/yd^3/d	kg BOD/m^3/d	0.5932
irrigation rate	gal/ft^2/h	m^3/m^2/h	0.049
specific surface area	ft^2/ft^3	m^2/m^3	3.28
Activated-sludge plant:			
loading	lb BOD/1000 ft^3/d	kg BOD/m^3/d	0.016
sludge loading		g BOD/g activated sludge per day	
sludge-volume index		ml/g	
air	ft^3/gal	m^3/m^3	6.23
	ft^3/lb BOD removed	m^3/kg BOD removed	0.0624
	ft^3/min ft^2 plan area	m^3/s m^2 plan area	0.00508
air pressure	lb/in^2	bar	0.06895
air compressor rating	ft^3/min	m^3/s	0.000472
power	hp/mil gal/d	W/m^3/d	0.164
Tertiary treatment			
Sand filters:			
loading	gal/ft^2/h	m^3/m^2/d	1.176
backwashing rate	gal/ft^2/min	m^3/m^2/d	70.56
Grass plots	mil gal/acre/d	m^3/m^2/d	1.123
Sludge treatment			
Sludge:			
volume	mil gal	m^3	4546
production	gal/hd	m^3/hd	0.004546
	lb dry solids/hd/d	kg dry solids/hd/d	0.4536
	lb volatile matter/hd/d	kg volatile matter/hd/d	0.4536
calorific value	Btu/lb	kJ/kg	2.326
specific resistance	s^2/g at 500 g/cm^2	m/kg at 49 kPa	9.81 × 10^3
Sludge gas:			
production	ft^3/lb volatile matter destroyed	m^3/kg volatile matter destroyed	0.0624
calorific value	Btu/ft^3	kJ/m^3	37.26
Sludge dewatering:			
yield	lb dry solids/ft^2/h	kg/m^2/h	4.881
Sludge digestion:			
loading	lb volatile matter/ft^3/d	kg/m^3/d	16.03
Sludge heaters:			
rating	10^6 Btu/h	MW	0.2931
gas consumption	ft^3/h	m^3/h	0.02832
Fuel oil:			
calorific value	Btu/gal	kJ/l	0.2321
	Btu/lb	kJ/kg	2.326

Index

Acinetobacter Calcoaceticus, 193
activated sludge, 93–100
   aeration methods, 94
   anoxic zones, 98–9
   bulking, 98
   characteristics, 78–80
   design parameters, 90–97, 105–6
   DO control, 94–6
   MLSS levels, 93
   operating problems, 78, 97–8
   shock loads, 99–100
alkaline cleaners, 55, 57
analysis, 230–38
   by GC/MS, 213
   for BOD, 231, 233–4
   for COD, 234
   for DO, 232
   for NH_3, 235–7
   for NO_3, 237
   for pH, 237–8
   for SS, 235
analytical suites, 232–3
anodising, 57, 64

BAF *see* submerged aerated filters
Bathing Water Quality Directive, 3, 202
BIMSch V90 emission standard 1990, 146
Biobead treatment system, 183–4
Biocarbone treatment system, 181–3
biofilters
   conventional, 82–7
   high rate, 87–90
   maintenance, 85
   operating parameters, 104–5
   operating problems, 79, 84, 86, 90
biological treatment, 73–82
   carbonaceous oxidation, 73, 80–81
   characteristics, 75–82

design parameters, 78–80, 104–7
microbiology, 80–82
nitrification, 73, 80–81
biologically aerated filters (BAF) *see*
   submerged aerated filters (SAF)
Biopaq treatment system, 187
Biostyr treatment system, 181
Biothane treatment system, 187–8
Black List (UK) substances, 176

cadmium effluent treatment, 55, 61–2, 63–4
carbonaceous oxidation, 73, 81
cesspools 151–7
   commissioning, 161–2
   costs, 160
   installation, 157
   layout/operation/maintenance, 163–5
Chemelec cell, 66–8
CHP *see* Combined Heat and Power
chromium effluent treatment, 55, 62
Combined Heat and Power (CHP), 137–8
continuous monitoring, 228–30
Copasack, 5–6
copper effluent treatment, 61–2
COSHH, 240
cyanide effluent treatment, 62–3

DAF *see* dissolved air flotation
dairy effluent case history, 242–6
deep shaft treatment system, 190–92
detritus removal methods, 1–16
dewatering methods for sludges, 119–33
digestion *see* sludge digestion
dissolved air flotation (DAF), 43–53
   application to chemical industries, 48
   application to dairy/food wastewaters, 45, 47, 242–6
   loading rates, 43–5

dissolved air flotation (DAF) (*continued*)
  operating parameters, 52–3
  operating problems, 51
  plant layout, 46

EEC Directive 76/160/EEC (1976), 3, 202
EEC Directive 85/337/EEC (1985), 147
EEC Directive 86/278/EEC (1986), 115
EEC Directive 89/369/EEC (1989), 145–6
EEC Directive 91/271/EEC (1991), 3, 115, 192, 202, 241
EEC Directive 91/338/EEC (1991), 56
EEC Directive 91/676/EEC (1991), 115
EEC Directive 91/689/EEC (1991), 115–16
effluent Consent standards, 74–5
emission standards (incinerators) Directive, 146
environmental impact assessment Directive, 147
Environmental Protection Act (1990), 15, 19, 146
environmental protection (controls on injurious substances) (No. 2) regulations 1993, 56
extended aeration, 100–103
  operating parameters, 100–101, 106–7
  operating problems, 101–2
  MLLS levels, 102

filter sacks, 1, 5–6
fines, 41
flotation *see* dissolved air flotation
flow measurement, 224–6
food manufacturing effluent case history, 248–50

gas emission standards, 146–8
genetic engineering, 204–5
grease traps, 5, 155, 162–3
grit composition, 17, 19
grit removal methods, 16–19
  by channels, 17, 19
  by detritor, 17, 18
  by vortex separator, 17, 20

Hazard Waste Directive, 115
HMSO/SCA analytical methods, 231

ion exchange, 65–6

lagoons, 33–4
lamella plate separators, 199–201
landfill levy, 147

membrane filtration, 201–3
metal ion biosorption, 70–71
metal plating effluent treatment, 62–4
  case history, 246–8
  Consent discharge limits, 56–9
  containing cadmium, 52, 61–4
  containing chromium, 55, 62
  containing cyanide, 62–3
  containing zinc, nickel, copper, 61–2
  from anodising, 64
  housekeeping, 60–61
  ion exchange, 65–6
  metallic particles, 65
  optimum pH values, 61–2
  photo-etching waste, 65
  pickling solutions, 64
  plating shop layout, 68–70
  toxicity, 56–9
metal recovery by electrolysis, 66–9
metals
  degreasing, 56
  fixing in sludges, 113
  inhibiting digestion, 59
  levels in incinerator ash, 145
  toxic to sewage treatment, 58
  toxic to trout, 58
mineral acids, 55–6
Mogden Formula, 199–200

NAMAS, 230, 249
Nitrates Directive, 115
North Sea ministerial conference 1991, 113

odour sources
  biological treatment systems, 75–6, 78, 90
  cesspools, 155
  fat and grease, 5
  grit, 17
  septic tanks, 155, 165
  sludge treatment, 27, 112, 115, 122, 124, 173
  soakaways, 167

oil discharges, 4
oil/fuel separation, 19–23
oil/water separation, 21–2, 48–9
on-line sensors, 228–30
oxidation ditches *see* extended aeration

package plant, 152–4, 158–63, 168–72, 179–80, 246–8
phosphorus removal in effluents, 192–5
Phragmites Australis, 188–9
Psychoda fly, 90

rag-balling, 2
Red List (UK) substances, 204
reed beds, 188–9
root zone treatment system, 188–9
rotating biological contactors (RBC), 90–92
   case history, 250–54
   commissioning, 162
   costs, 161
   installation, 158–9
   layout/operation/maintenance, 168–72
   loadings, 91
   operating parameters, 105
Royal Commission on Environmental Pollution 17th Report, 116

SAF *see* Submerged Aerated Filters
sampling, 207–20
   by automatic samplers, 217–20
   industrial sites, 215–20
   manual, 216–17
   objectives, 220–21
   sewage works, 212–15
   statistics, 220–24
sampling methods, 209–12
   continuous, 209–10
   flow proportional, 210–11
   manual, 211–12
   time average, 211
screenings
   bagging, 13
   dewatering/compaction, 13–14
   discharges to sea, 3
   disposal methods, 12–16
   incineration, 15–16

   maceration, 12–13
   plastic, 3, 6, 39
screens
   coarse bar, 9–10
   fine, 11–12
   headloss, 10
   manual, 4–8
   mechanical, 8–16
   rundown, 7–8
septic tanks, 151–8
   commissioning, 161–2
   costs, 160
   installation, 154, 157–8
   layout/operation/maintenance, 165–7
septic tank soakaways, 152, 156
   costs, 160
   installation, 154, 157–8
   layout/operation/maintenance, 167–8
   soil porosity, 158
settlement
   methods, 25–41
   performance for bacterial removal, 26
   performance for BOD/COD removal, 25–6
   performance for SS removal, 25–6
   theory, 28–9
settlement tanks, 27–51
   circular, 35
   design parameters, 29–33, 35, 52
   desludging, 27, 39–40
   maximum solids loading, 31–3
   operating problems, 49–51
   rectangular, 34
   retention times, 29–31
   scraper mechanisms, 34–40
   scum collection, 38–9, 49
   upward flow rate, 29, 41–3, 49
sludge
   ash, 144–5
   blanket, 42
   characteristics
      industrial, 118–19
      inorganic, 111
      organic, 112
      sewage sludge, 112, 116–17
   density, 26
   dry matter, 27
   nitrogen content, 27
   stirred specific volume (SSV), 32–3

sludge digestion
  anaerobic heated, 135–40
  cold, 135
  costs, 140
  farm slurry, 138–9
  gas, 134, 137–8
  inhibition, 58–9, 139
  microbiology, 133–4
  operating characteristics, 135–40
sludge treatment methods
  belt press, 127–8
  centrifuge, 128–30
  composting, 121–2
  dewatering, 113, 118–19, 122–33
  digestion *see* sludge digestion
  drying, 130–33
  incineration, 140–47
  picket fence thickener, 121
  plate press, 124–7
  screw press, 128–9
  sewage sludge, 115–17
Sludge Use in Agriculture Regulations 1989, 115
small sewage treatment plant, 152–6, 158–62, 163 *see also* package plant
  commissioning, 162
  costs, 160–61
  installation, 158–60
  layout/operation/maintenance, 172–8
Stokes Law, 28–9
Submerged Aerated Filters (SAF), 180–84
  bacterial removal rates, 183
  biological performance, 181
  hydraulic loading, 181, 183
  power consumption, 181, 183

TA Luft emission standard 1986, 146
Thames flooded filter, 183
tilted plate oil separator, 21, 23
toxic discharges
  effect on activated sludge, 99–100
  toxic metals, 54, 58–9
trade effluent charging, 226–8
trade effluent Consents, 4

UASB *see* upward flow anaerobic sludge blanket treatment system
ultra violet (UV) disinfection, 195–8
  bacterial/viral sensitivity, 196
  transmission values for wastewaters, 197
upward flow anaerobic sludge blanket treatment system (UASB), 185–8
upward flow clarifiers, 41–4
  fines, 41
  flow rates, 42–3
  operating parameters, 52
  operating problems, 51
  sludge blanket, 42
Urban Waste Water Directive (UWWD), 3, 115, 202, 241

Water Industry Act 1991, 228
wedge wire panel, 41–2, 49
wet oxidation processes, 198–9

zoogloeal slime, 81–2